SpringerBriefs in Applied Sciences and Technology

Safety Management

Series Editors

Eric Marsden, FonCSI, Toulouse, France

Caroline Kamaté, FonCSI, Toulouse, France

Corinne Bieder, FonCSI, Toulouse, France

The *SpringerBriefs in Safety Management* present cutting-edge research results on the management of technological risks and decision-making in high-stakes settings.

Decision-making in high-hazard environments is often affected by uncertainty and ambiguity; it is characterized by trade-offs between multiple, competing objectives. Managers and regulators need conceptual tools to help them develop risk management strategies, establish appropriate compromises and justify their decisions in such ambiguous settings. This series weaves together insights from multiple scientific disciplines that shed light on these problems, including organization studies, psychology, sociology, economics, law and engineering. It explores novel topics related to safety management, anticipating operational challenges in high-hazard industries and the societal concerns associated with these activities.

These publications are by and for academics and practitioners (industry, regulators) in safety management and risk research. Relevant industry sectors include nuclear, offshore oil and gas, chemicals processing, aviation, railways, construction and healthcare. Some emphasis is placed on explaining concepts to a non-specialized audience, and the shorter format ensures a concentrated approach to the topics treated.

The *SpringerBriefs in Safety Management* series is coordinated by the Foundation for an Industrial Safety Culture (FonCSI), a public-interest research foundation based in Toulouse, France. The FonCSI funds research on industrial safety and the management of technological risks, identifies and highlights new ideas and innovative practices, and disseminates research results to all interested parties. For more information: https://www.foncsi.org/

Mathilde Bourrier
Editor

Decommissioning Aging Installations and Declining Technologies

Burden or Inspiration?

Editor
Mathilde Bourrier
Department of Sociology
University of Geneva
Geneva, Switzerland

ISSN 2191-530X ISSN 2191-5318 (electronic)
SpringerBriefs in Applied Sciences and Technology
ISSN 2520-8004 ISSN 2520-8012 (electronic)
SpringerBriefs in Safety Management
ISBN 978-3-031-88368-2 ISBN 978-3-031-88369-9 (eBook)
https://doi.org/10.1007/978-3-031-88369-9

This work was supported by Fondation pour une Culture de Sécurité Industrielle.

© The Editor(s) (if applicable) and The Author(s) 2025. This book is an open access publication.

Open Access This book is licensed under the terms of the Creative Commons Attribution 4.0 International License (http://creativecommons.org/licenses/by/4.0/), which permits use, sharing, adaptation, distribution and reproduction in any medium or format, as long as you give appropriate credit to the original author(s) and the source, provide a link to the Creative Commons license and indicate if changes were made.
The images or other third party material in this book are included in the book's Creative Commons license, unless indicated otherwise in a credit line to the material. If material is not included in the book's Creative Commons license and your intended use is not permitted by statutory regulation or exceeds the permitted use, you will need to obtain permission directly from the copyright holder.
The use of general descriptive names, registered names, trademarks, service marks, etc. in this publication does not imply, even in the absence of a specific statement, that such names are exempt from the relevant protective laws and regulations and therefore free for general use.
The publisher, the authors and the editors are safe to assume that the advice and information in this book are believed to be true and accurate at the date of publication. Neither the publisher nor the authors or the editors give a warranty, expressed or implied, with respect to the material contained herein or for any errors or omissions that may have been made. The publisher remains neutral with regard to jurisdictional claims in published maps and institutional affiliations.

This Springer imprint is published by the registered company Springer Nature Switzerland AG
The registered company address is: Gewerbestrasse 11, 6330 Cham, Switzerland

If disposing of this product, please recycle the paper.

Series Editor's Foreword

This collective volume is the fruit of a workshop organized by New Technology and Work (NeTWork), an international, interdisciplinary group of academics, regulators and practitioners. NeTWork aims to provide concepts and methods for addressing individual, organizational and societal risks created by technological development, for evaluating the state of the art of technology management, regulation and risk control and debating the way forward. Since 1983, NeTWork has held annual workshops relating to the overall theme of new technologies and work. Workshops have covered a wide range of topics that included human error, accident investigation, training, distributed decision-making and management. Recent preoccupations have focused more specifically on themes of great scientific and social significance: the safety of technology-intensive systems and the role of human contribution to either failure or resilience of high-hazard activities or the interplay of climate change and safety in high-risk industries.

The Foundation for an Industrial Safety Culture (FonCSI) is a public-interest research foundation that brings together representatives mainly from hazardous industries, regulatory bodies, local authorities and the international academic community. Operating as an idea lab, FonCSI questions the organizational, inter-organizational and societal dimensions of the management and governance of safety in context, uncovering new paradigms and practices. FonCSI is proud to have been supporting and contributing to the activities of NeTWork since 2010.

This volume addresses the issue of decommissioning ageing installations and declining technologies. Although mainly overlooked so far as a domain by safety scientists, the theoretical and empirical contributions from a diversity of experts unravelled the richness of this usually discreet life-phase of technology. Yet, it turns out that the end of lifecycle of technologies concentrate questions and issues that are central to safety and can be valuable sources of inspiration for safety management and governance. Many thanks to the godmother and godfather who organized this workshop, Mathilde Bourrier and Eric Marsden, and to all the contributors for their thought-provoking contributions.

On a more sombre note, we learned early in 2025 of the sudden death of one of the participants in the workshop that led to the production of this collective volume, Nicolas Nova. His energy and his enthusiasm in analysing the partly subterranean practices of passionate digital archivists and recyclers were an inspiration to the workshop participants.

February 2025

Corinne Bieder
FonCSI
Toulouse, France

Contents

1. **Decommissioning Ageing Installations and Declining Technologies: Burden or Inspiration?** 1
   Mathilde Bourrier and Eric Marsden

2. **Technological Decline from a Socio-Material Perspective** 9
   Zahar Koretsky

3. **Discontinuation Through Enforcement of the Law: Court Rulings as Leverage for Stopping Delegitimized Practices and Technologies** 21
   Peter Stegmaier

4. **Undesirable Systems, Undesired Ends: The (Un)Bearable Heaviness of Phasing-Out Pesticides?** 35
   Bruno Turnheim, Marc Barbier, and Mireille Matt

5. **Glory, Mourning, Memory. Archiving Knowledge, Dismantling Nuclear Power** 49
   Christine Bergé

6. **Preserving and Valuing Memory for a Safer and More Sustainable Future: The Key Role of Archives** 61
   Fabienne Peris-Raimbault

7. **Collecting, Dismantling, Documenting, Reusing: Marginal Practices with Discarded Electronics** 73
   Nicolas Nova, Anaïs Bloch, and Thibault Le Page

8. **Safety Culture Lessons Learned in Decommissioning VTT's FiR 1 Research Reactor** 83
   Kaupo Viitanen, Merja Airola, Markus Airila, and Petri Kotiluoto

9 **Decommissioning Management and Leadership for Safety Education: Addressing the Organizational Challenges and the Managerial Complexity of Nuclear Decommissioning Projects** 95
Yoann Guntzburger, Jacques Repussard, Savéria Cecchi, Pierre Daniel, Renata Kaminska, Joseph A. Ridao Cabrerizo, Evelyne Rouby, and Catherine Thomas

10 **The Language of Transitions: Navigating Innovation, Decline and Renewal** 107
Eric Marsden and Mathilde Bourrier

Chapter 1
Decommissioning Ageing Installations and Declining Technologies: Burden or Inspiration?

Mathilde Bourrier and Eric Marsden

Abstract This book addresses the question of ageing and declining technologies, induced or not, and bets on the fact, that there is much to gain in looking closely at the activities entailed. Dismantling, decommissioning, deconstruction, closing-down, phasing-out, discontinuation, redirecting are all operations that are key to numerous industry sectors and high-hazard activities. Why? Because they represent a form of undoing that may prove to be key in redirecting vast industrial sectors, in light of climate change urgent transition. These operations are rarely the most "glamorous" phase of a system's lifecycle, and may even constitute a form of taboo in some industry sectors. Depollution and waste management are usually envisioned as a *mal necessaire*, rather than as an expected and intentional industrial phase, from which may emerge precious knowledge, expertise and know-how. This book proposes a different approach and a side step, considering that there are different ways to go about decline. The contributors, drawn mainly from academic backgrounds, embarked on the idea that there is much to learn from these phases, also because as the chapters will show, they are replete with surprises, treasures, uncertainties, and reversals. This eight-chapter academic book is based on presentations and earlier drafts by international experts who were invited in January 2024 to the 37th workshop under the auspices of NetWork (New technologies and Work), supported by FonCSI, a French public-interest research foundation. It deals with issues of interest to researchers and graduate students in safety science, transition studies, science and technology studies, organization and management studies, and anthropology, as well as to members of expert bodies and experts in industry and consultancy concerned with similar subjects.

Keywords Dismantling · Decommissioning · Deconstruction · Closing-down · Phasing-out · Discontinuation · Decline · Destabilization

M. Bourrier (✉)
University of Geneva, Geneva, Switzerland
e-mail: Mathilde.Bourrier@unige.ch

E. Marsden
FonCSI, Toulouse, France

1.1 Background and Opening Questions

Dismantling, decommissioning, deconstruction, closing-down, phasing-out, discontinuation, redirecting are all operations that are key to numerous industry sectors and high-hazard activities. Despite their importance, depollution and waste management are not the most prestigious categories of industrial activity, and they tend to be geographically and socially hidden away from public view or sometimes used to contest an industry (Cotton 2022; Schöbel et al. 2017). For example, high-level nuclear waste repositories already have a long history paved with setbacks, social unrest, social, and technological controversies (Macfarlane 2003; Macfarlane and Ewing 2006). These projects could be a source of inspiration (Schläppy et al. 2021). Yet, often, their stories are untold because ageing facilities, ageing and declining technologies, and zombie technologies are the hidden face of a techno-sphere that everyone would prefer to ignore and know less about than more (Bergé 2010 and 2011).

The decommissioning and phasing-out of ageing installations does entail specific safety risks, first for the workers but also ultimately for the surrounding population (Ashforth 1999). There is also a transnational dimension to include, with the offshoring of dismantling (in Bangladesh for instance in the case of ships) associated with pollution and with health and safety problems (asbestos, gas explosions, water contamination). These outsourced industrial practices say something about the activities that we would rather hide in very poor countries. Some countries have made a specialty of these hazardous waste treatment operations (Tanha et al. 2022).

Culturally, professional identities in the engineering world are geared towards innovation, breakthroughs and the creation of new technologies and installations. It has been often noted that maintenance activities, integral to numerous industrial processes, would greatly benefit from better consideration in the earlier lifecycle phases of design and conception (Denis and Pontille 2015, 2025). It is often regretted that they lack the same amount of attention compared to the construction and operations phases. Ageing installations, or legacy facilities in palliative care, are often envisioned as a burden rather than as a source of genuine learning. From ageing to decommissioning across safety-critical systems, there is a lot to uncover (D'Agata 2011; O'Leary 2022).

Of course, this general argument needs to be refined. Some industries have developed dedicated supply chains and specific expertise on how to decommission nuclear power plants, treat radioactive waste and polluted soil, recycle construction materials, remove asbestos from buildings, etc. Specific know-how has been forged through both the "normal" and the catastrophic life of industrial sites (Chernobyl, Fukushima, Deepwater Horizon spill are of course in every mind). Interestingly, these operations are either largely invisible or kept from the view of the general public or worshipped as heroic technical prowess. In and for themselves, these activities are interesting to look at, despite their gloomy reputation. But they entail a much bigger interest in light of the current climate urgency (Bonnet et al. 2021; Ialenti, 2020).

Numerous industry sectors and their related infrastructures are already being (coal, oil and gas, certain nuclear facilities) or will soon need to be (internal combustion engines, kerosene-fueled aviation) dismantled, phased-out, closed down. It will take decades to be able to progressively detach ourselves from forms of industrial production that are integral to our subsistence, yet equally detrimental to further life on earth. There is indeed much more to learn from examples taken from these industrial phases.

The issue of phasing-out carries a lot of complex problems, technical, and safety wise of course, but also social, patrimonial, symbolic, psychological (de Wildt 2020; Joyce 2020; Laraia 2019; Wall 2020). Bringing an entire installation to an end is emotionally draining for employees who worked and devoted time, energy, and skills to its operations and maintenance. This "aftercare" phase (Stegmaier 2023; Goulet and Vinck 2023) is also seen as less appealing, less attractive professionally. While there is some literature on the technical dimensions of ageing facilities and equipment, there is far less work on the organizational and societal aspects (Martin and Guarnieri 2013).

There are many different points of entry in the subject. We have listed below some questions which were put together to stir the conversation at the NeTWork workshop. Are the lessons from existing decommissioning projects sufficiently fed into the design of new projects and the operation of existing ones? How to prepare for decommissioning during design and/or operation? Is this anticipation a regulatory requirement in different industry sectors? What are the tradeoffs to consider between designing for ease of dismantling and for safe operation? Recently, for economic and/or energy supply related reasons, decisions have been made to grant lifetime extensions instead of phasing-out ageing installations: what are the consequences of such decisions, both from a safety standpoint but also from the consideration given to this dismantling phase, endlessly postponed? What governance mechanisms for these tradeoffs? What does it entail from an engineering and organizational point of view? How to manage competencies and careers related to technologies with a long service life (there are concerns here in satellites, aviation, railways, nuclear power, and banking) or which may seem less inspiring and attractive professionally? Can the concept of restoration and renewal, which is implicit in decommissioning, appeal to younger generations of workers, whose environmental consciousness tends to be more developed? While some countries with less developed regulatory frameworks have become convenient hosts for the disposal (and in principle, reuse) of dangerous materials, ships, oil platforms, what are the possibilities to advocate for safer practices? How to capitalize upon and transfer the immense and varied know-how that decommissioning projects generate? How much is transferable from one project to another, from one sector to another? Intergenerational issues have also to be considered among workforces, linked with La Porte's concept of "Never-ending Management and Stewardship" (La Porte and Metlay 1996). What are the key issues from the angle of workers' occupational health and safety? How should society manage the burden of legacy technologies and industries and brownfield sites (e.g. mandatory environmental liability insurance, legal obligations concerning decommissioning and restoration funds, mandatory disclosure in firms' extra-financial reporting, targeted actions by safety authorities)? Are the lessons from past failures (coal and steel industry, Leboutte 2009), applied to new industries, such as offshore wind power?

However, as we shall see, not all of these questions have triggered contributions in the book. Several topics that we deemed important have been left unattended to, for now. This leads us to think that the topic should stay high on the agenda of researchers and programmes managers. With this book we hope to start a conversation on the conditions under which current examples of decommissioning and phasing-out across safety-critical systems could help establish a basis for envisioning future dismantling, in light of the sustainability transitions that many sectors have to engage with. Under what conditions can we make these operations more than a sad act of deconstruction, but a gesture to learn about technological rebound, renaissance, and ecological redirection? Under what conditions can these critical, often long-term operations, seen as wicked problems, represent a rich field of experimentation? What would it take to seize these moments as culturally rich, replete with patrimonial sensemaking? This book is only the beginning of a marked interest for activities that require vigilance and better understanding in order to ensure the collection of the largest portfolio of empirical cases possible.

1.2 What's in the Book?

Despite evident safety issues related to phasing-out industrial activities, it is of interest to note that the classical safety science community is not at the forefront of the current reflections. On the contrary, the vibrant community of science and technology studies and its offspring, transition studies, have so far made the most sustained effort to understand what's going on. Phase-out, destabilization, and decline have become a dedicated stream of research in transition studies (Koretsky et al. 2022). Three chapters in this book (Koretsky; Stegmaier; Turnheim, Barbier and Matt) offer a flavour of the richness of these approaches.

Koretsky introduces a conceptual framework linking three main components "materials", "competences", and "meanings" and the ways in which their relative alignments or misalignments contribute to decline of technologies. The framework is tested against a large data set of cases. Koretsky's chapter presents us with three cases of "weak" decline (incandescent light bulbs in the EU, cloud seeding in the USA, and the Ural computer in Russia) and documents how misalignments of one of the key components have led to specific pathways to decline.

Stegmaier adds the legal dimension to the crucial discussion of governing decline, where "doing discontinuation" can be observed and understood as governance-in-action. He presents three recent case studies: an inner-city ban on diesel vehicles in North Rhine-Westphalia in Germany, the strengthening of the Climate Change Act through the Federal Constitutional Court in Germany, and the decision in the case of Milieudefensie et al. v. Royal Dutch Shell plc. in the Netherlands. Stegmaier focuses on court proceedings that mobilize existing laws, and analyses how specific groups and actors are literally forcing discontinuation via the courts, what he calls "discontinuation through judicial action".

And finally, Turnheim and colleagues with a precise historical approach convince us that the long march and never-ending story of pesticide phase-out has to be envisioned as a systemic problem affecting the entire agricultural sector.

A second line of contributions (Bergé; Peris-Raimbault; Nova, Bloch and Le Page) come from the ranks of anthropology, archives science and museography realm, which can, at first, constitute a surprise but which, taking a deeper look, opens up the crucial discussion of documenting and preserving knowledge and know-how. The anthropologist Bergé compares the slow deconstruction of the Superphénix nuclear reactor to the deconstruction and refoundation of Lyon's Natural History Museum, which she calls institutions of knowledge and memory. These institutions harbour treasures of ancient technological, material, and artistic practices, whose preservation is essential if we are to continue to understand their embodied knowledge and social practices.

The archivist Peris-Raimbault discusses aeronautics industry archives as an example, introducing the reader to the constant renewed practice of industrial archiving as a driving force for innovation and transmission of knowledge in a perspective of industrial safety and sustainability.

Anthropologist Nova and colleagues Bloch and Le Page have produced a visual essay that takes a close-up look at amateur practices for preserving and diverting obsolete information and communication technology objects (video game consoles, old video games, old computers, all kinds of computer waste...). These little-known, often informal practices, conceal a treasure trove of heritage, playful or scholarly intentions that could inspire the vast technological deconstruction projects currently underway.

A third angle is represented in the book by the chapters of Viitanen and Guntzburger and their respective colleagues. Anchored in the management and safety science, they both tackle the issue of decommissioning and dismantling nuclear facilities. Both chapters note that there already exists a wealth of regulatory texts and guidance pertaining to such complex projects. However, these efforts are not matched with a similar managerial doctrine capable of handling such large-scale projects in the long run. The decommissioning and deconstruction projects are very different from the construction ones. They require a different skill set and perhaps also a different mindset. Too often, a lack of proper education and training curriculum is missing. Developing a safety culture framework tailored to the specific needs of decommissioning is not a mere variation of a regular safety culture framework used during normal operations.

The work is only beginning. There is an urgent need to envision legacy industries, ageing and declining technologies, in need of phasing-out, deconstruction, decommissioning, dismantling, as witnesses from a past, that it is impossible to get rid of without proper care. Special care is needed to ensure that patrimonial, anthropological, historical, material, sociological, and technological dimensions all concur to resituate the proper place that such technologies played in human history. An urgent de-consecration is under way and requires proper documentation and accurate storytelling.

References

B.E. Ashforth, G.E. Kreiner, "How can you do it?": Dirty work and the challenge of constructing a positive identity. Acad. Manag. Rev. **24**, 413–434 (1999)

C. Bergé, *Superphénix, déconstruction d'un mythe* (La Découverte, Paris, 2010)

C. Bergé, Slow death of the Superphénix, Le Monde Diplomatique (2011)

E. Bonnet, D. Landivar, A. Monnin, Héritage et fermeture : une écologie du démantèlement. Éditions divergences (2021)

M. Cotton, Deep borehole disposal of nuclear waste: trust, cost and social acceptability. J. Risk Res. **25**(5), 632–647 (2022)

J. D'Agata, About a mountain (WW Norton & Company, 2011)

J. Denis, D. Pontille, Material ordering and the care of things. Sci. Technol. Hum. Values **40**(3), 338–367 (2015)

J. Denis, D. Pontille, The care of things: ethics and politics of maintenance (John Wiley & Sons, 2025)

F. Goulet, D. Vinck (eds.), New horizons for innovation studies: doing without, doing with less (Edward Elgar Publishing, 2023)

V. Ialenti, Deep time reckoning: how future thinking can help Earth now (MIT Press, 2020)

R. Joyce, The future of nuclear waste: what art and archaeology can tell us about securing the world's most hazardous material (Oxford University Press, 2020)

Z. Koretsky, et al., Technologies in decline: socio-technical approaches to discontinuation and destabilisation (Taylor & Francis, 2022)

M. Laraia, Beyond decommissioning: the reuse and redevelopment of nuclear installations (Woodhead Publishing, 2019)

R. Leboutte, A space of European de-industrialisation in the late twentieth century: Nord/Pas-de-Calais, Wallonia and the Ruhrgebiet. European Review of History—Revue européenne d'histoire **16**(5), 755–770 (2009)

A. Macfarlane, Underlying Yucca Mountain: the interplay of geology and policy in nuclear waste disposal. Soc. Stud. Sci. **33**(5), 783–807 (2003)

A. Macfarlane, R. C. Ewing, Uncertainty underground: Yucca Mountain and the nation's high-level nuclear waste (MIT Press, Cambridge, 2006)

C. Martin, F. Guarnieri, Nuclear decommissioning and organizational reliability: involving subcontractors in collective action. in *Decommissioning Challenges: An Industrial Reality and Prospects, 5th International Conference*, vol. 9 (Avignon, France, 2013), p. hal-00815982

D.W. O'Leary, Why aren't we talking about nuclear waste?. New Atl. **67**, 40–43 (2022)

T.R. La Porte, D.S. Metlay, Hazards and institutional trustworthiness: facing a deficit of trust. Public Adm. Rev. 341–347 (1996)

M-L. Schläppy, et al., Trash or treasure? Considerations for future ecological research to inform oil and gas decommissioning. Front. Mar. Sci. **8**, 642539 (2021)

M. Schöbel, R. Hertwig, J. Rieskamp, Phasing out a risky technology: an endgame problem in German nuclear power plants? Behav. Sci. Policy **3**(2), 41–54 (2017)

P. Stegmaier, Aftercare, or doing less with discontinuation niche governance, in *New Horizons for Innovation Studies,* (Edward Elgar Publishing, UK, 2023). https://doi.org/10.4337/9781803925554.00029.

M. Tanha, G. Michelson, M. Chowdhury, P. Castka, Shipbreaking in Bangladesh: organizational responses, ethics, and varieties of employee safety. J. Safety Res. **80**, 14–26 (2022)

W. Wall Finnish-Swedish television drama, ed. by A. Salmenperä, A. Zackrisson, M. Pöllä, R. Lehtinen (2020)

K. de Wildt, Ritual void or ritual muddle? Deconsecration rites of Roman Catholic Church Buildings. Religions **11**(10), 517 (2020)

Open Access This chapter is licensed under the terms of the Creative Commons Attribution 4.0 International License (http://creativecommons.org/licenses/by/4.0/), which permits use, sharing, adaptation, distribution and reproduction in any medium or format, as long as you give appropriate credit to the original author(s) and the source, provide a link to the Creative Commons license and indicate if changes were made.

The images or other third party material in this chapter are included in the chapter's Creative Commons license, unless indicated otherwise in a credit line to the material. If material is not included in the chapter's Creative Commons license and your intended use is not permitted by statutory regulation or exceeds the permitted use, you will need to obtain permission directly from the copyright holder.

Chapter 2
Technological Decline from a Socio-Material Perspective

Zahar Koretsky

Abstract Safety and risk assessment and management involve, among other, anticipating, preventing, and planning for technological decline, which may come in the form of ageing and disuse of a technological system, or its coordinated phase-out. In the interdisciplinary literature on sustainability transitions, coordinated phase-out is typically the centre of interest. Technological decline may be understood as the process of diminishing production and/or use of a given technology. Using three empirical examples, the chapter illustrates and identifies two types of decline: weak and strong. Weak decline involves reversible decline and lower costs of re-emergence, whereas strong decline signifies high irreversibility. The chapter concludes with practical recommendations for risk assessment and safety management, emphasizing the need for strategic mapping of technologies and coordinated efforts from various stakeholders.

Keywords Decline · Phase-out · Irreversibility

2.1 Introduction

Safety and risk assessment and management involve, among other, anticipating, preventing, and planning for technological decline, which may come in the form of ageing and disuse of a technological system or its coordinated phase-out. Compliance with health, safety, and environmental protection regulations is a key concern here. In the interdisciplinary literature on sustainability transitions, coordinated phase-out is typically the centre of interest. Here questions like "under what conditions is the phasing-out of a technology both feasible and irreversible?" are asked. Technological decline may be seen as a broad class of phenomena that encompasses ageing and disuse as well as coordinated phase-out, traceable by diminishing production and/or use of the technology in question. It is also a complex phenomenon, often involving active confrontation with powerful stakeholders and the management of stranded

Z. Koretsky (✉)
Utrecht University, Utrecht, The Netherlands
e-mail: zahar.koretsky@pm.me

© The Author(s) 2025
M. Bourrier (ed.), *Decommissioning Aging Installations and Declining Technologies*,
SpringerBriefs in Safety Management, https://doi.org/10.1007/978-3-031-88369-9_2

assets, emotional attachments of individuals, and impacts on displaced workers (e.g. from closed coal mines). There is also uncertainty regarding the lasting effect of technology decline and its wider implications.

This chapter is aimed to contribute to building bridges between sustainability transitions studies and safety and risk assessment and management. The discourse in the sustainability transitions field on this subject is vibrant, with ongoing debates exploring terms like phase-out, destabilization, discontinuation, and exnovation (Sovacool et al. 2023; Koretsky et al. 2023), each emphasizing different facets of the phenomenon in question. In this chapter, I offer a discussion of technological decline for the safety and risk assessment and management community, bringing some of the recent developments in the sustainability transitions field.

I will start with a presentation of a theoretical framework that connects different concepts in order to put forward an explanation of how technologies and technological systems decline. I will then use three empirical examples of decline as an illustration and clarification of the framework. I conclude with practical implications.

2.2 Theoretical Framework for Analysing Decline from a Socio-material Perspective

The below theoretical framework has been created with the process of middle-range theory-building (Eisenhardt and Graebner 2007), i.e. I synthesized it with the help of literature review, rich empirical data analysis, and generation and, crucially, revision of theory.

To unpack the processes of coordinated and uncoordinated technological decline that, I propose, are productive to shed light on, I will discuss in brief the following concepts: technology, configuration and its components, and processes of alignment and misalignment. Together they will serve as a theoretical framework.

Technology. Literature on science and technology studies (STS), which sustainability transitions studies draw from and overlap with, takes an inclusive view on "technology" as any class of human-created objects: from stone knife, to clothing, to radio, to a factory, to the energy grid. Where other views tend to treat any given technology as a neutral tool (e.g. in engineering) or an efficiency factor (e.g. in economics), STS and sustainability transitions studies investigate technology as a sociological phenomenon, and study the societal interactions and inherent political dimensions of decisions concerning technology (Foucault 1977; Latour 1991). A key research interest here is to understand society better through the study of technology and vice versa.

Literature on the actor-network theory, security studies, feminist studies, as well as some of social practice theory converges on a socio-material understanding of technology. From this socio-material perspective, a given technology is viewed as a configuration of physical artefacts and social practices, where the design, functionality, and use of the technology are shaped by and, in turn, shape the social contexts in which it exists.

Technology as a configuration. I propose that a socio-material conceptualization of technology offers analytical advantages when studying ongoing and declining technology. By analytically untangling the relations, the analyst is able to describe linkages, their dynamics and patterns, and how alternatives are made less possible. Here it is useful to identify a unit of analysis—the key focal entity of study—and a unit of observation—entities to be studied or measured (Sheppard 2020). From a socio-material perspective, I propose to specify three units of analysis:

- materials, including such units of observation as objects, tools, hardware, physical infrastructure, production facilities, and resources;
- competences, including such units of observation as tacit or codified knowledge in design, production, or use; and
- meanings, including such units of observation as individual interpretations and emotions, and collective discourses, institutions, rules, and narratives.

For the purposes of analysis, I will treat these units of analysis as three types of components of a configuration that constitutes a technology. The material component provides the physical basis of technology, competences enable its production and use by actors who are the carriers of competences, and meanings reflect the actors' constructed needs and societal rules for the technology's existence—arguably the most crucial component, as seen in empirical cases e.g. (Koretsky 2022a).

Each component is a network of entities that coexist and compete. The dynamics within the components can be expected to be driven by the same rules as other networks, e.g. they are dynamically stable as long as there is alignment (Goulet 2021), and they can become unstable if there is too much internal contestation.

Alignment and misalignment. The three components can be aligned or misaligned. The word "alignment" is widely used ("ensure alignment with policy", "aligned their visions", etc.), but Goulet (Goulet 2021) proposed to specify it in the context of sustainability transitions studies and STS. In this chapter, I follow his qualification of alignments as "pairings between entities of the same nature, positioned in the niche and in the regime [and whose] strength … relies on the coherence of the new sets [of entities] formed" (Goulet 2021). In other words, alignment refers to actors pairing a component (resources, knowledge, narratives, etc.) that is (still) marginal, with one that is dominant. In the language of sustainability transitions studies, alignments ensure path dependency and lock-in (Goulet 2021). The inverse processes are, then, misalignments. Alignment may take the form of e.g. new projects funded or the appearance of positive framings in the media, while misalignment may look like e.g. a national sales ban.

In parallel with institutional theory, where continual reproduction is seen as key for continuity (Lawrence 1999), the reproduction and alignment of the three components are essential for maintaining the configuration. The reproduction and alignment of components in time and in different locations (e.g. regular use of gas turbine technology in specific plants in Germany or China) indicates and, to some extent, ensures continued production and use of technology. Without even one of the components (e.g. the competences to operate or maintain the gas turbine), the configuration could eventually collapse and begin declining. Reproduction is supported by legal, normative, institutional, and structural factors—collectively known as a "regime" in sustainability transitions studies (Geels 2007). I provide examples in the next section, where I illustrate the presented theoretical framework with three empirical cases of decline. All cases confine their analysis primarily to national borders.

2.3 Three Cases of Technological Decline

My cases are: incandescent light bulbs in the EU (Koretsky 2021), cloud seeding in the USA (Koretsky and Lente 2020), and the *Ural* computer in Russia (Koretsky et al. 2022). To select my cases, I looked primarily for observable markers of technological decline in the form of diminishing rates or units of production, consumption, or sale. In analysing the cases, I employed methods from historical sociology, specifically narrative analysis techniques, to construct causally linked storylines beyond mere chronological documentation of events. Following this method, I organized empirical data into narratives—from technology development to measured decline—which facilitates historical comparison.

To guide data collection, I established two retention criteria: (1) references to the technology and its synonyms, and (2) details identified during close reading of the sources, related to:

(a) the materials component: design characteristics, location and number of production facilities, funding amounts and sources, production, and use statistics;
(b) the competencies component: state of the art and its dynamics, necessary knowledge and skills and their dynamics, career changes of key individuals;
(c) the meanings component: personal accounts of key events from key people such as lead designer, (former) employee, or representative of government as a key user.

I gathered qualitative (textual quotes) and quantitative (statistics, bibliometrics) data from primary and secondary sources. Data were coded by year and unit of analysis (i.e. materials, competences, meanings), forming a database (Koretsky 2022b). Timeline visualization in Excel aided pattern analysis, highlighting key event chains (e.g. funding new projects, media framings) and historical phases (e.g. development, decline). This comprehensive approach, spanning around six months per case, enabled iterative analysis to discern emergent patterns.

2.3.1 Case 1: Cloud Seeding

Cloud seeding is a technique aimed at modifying atmospheric conditions to stimulate or suppress precipitation by introducing chemical compounds, typically dry ice or silver iodide, which can create ice crystals from air moisture. These compounds are dispersed from aircraft equipped with burners, flares, hoppers, or other delivery systems mounted on the wings or fuselage. The process of cloud seeding is often accompanied by photo cameras, water gauges or sensors, and meteorological models.

Cloud seeding projects emerged in the USA, USSR, and UK shortly after World War II. In the USA, cloud seeding experiments during the 1950s and 1960s were stimulated by a series of droughts which brought water supply control high on the public agenda. Cloud seeding was also explored by the military, including, as the press exposed, during the Vietnam War, which led to a national and international scandal. An international anti-"weather-modification" treaty was signed in 1977. Both the treaty and the outcry preceded the peaking and then dropping of cloud seeding funding and the closing-down of many cloud seeding projects in the USA. This is when the alignment between the three components started unravelling: a strong "counter"-meaning ("unacceptable ethical and environmental hazard") emerged and damaged the legitimacy of other meanings (such as "water supply control" and "forest fire relief"); as a result of lost legitimacy, the amount of government funding decreased, misaligning the materials component. This misalignment threatened the existence of the configuration as a whole.

There was a long pause in publicly visible activity related to cloud seeding, during which the field of meteorology emerged and made use of the models developed for cloud seeding projects. But in the early 2010s, cloud seeding appeared in the US congressional debates on geoengineering, driven by scientists proposing it as a solution for global warming. These efforts enabled alignment of the meanings component, as interested actors linked the new meaning of geoengineering with old and new materials and competences, ultimately revitalizing, re-aligning the configuration of cloud seeding technology.

In sum, the history of decline and revival of cloud seeding was marked by these key events:

(a) the formulation of a critical narrative by the press, amplified by Congress and civil society, who were already agitated by the pacifist and environmental movements;
(b) governance decisions to withdraw funding and endorse a 1977 international anti-cloud-seeding treaty; and
(c) the revival of academic and political discourse around global warming, positioning cloud seeding as a potential response.

In the end, cloud seeding is a case of incomplete and reversed decline of the configuration. It shows that viewing technology as a configuration of socio-material components that constitute it can be productive. Using this theoretical lens, I am able to illustrate that a misalignment between the components seems to coincide with a decrease in diversity inside of one or more components—in this case, meanings. In other words, a technology starts to decline when its application becomes more narrowly perceived.[1] Additionally, during the misalignment of the configuration, some forms of competences seem to travel to other configurations and technology fields (e.g. from cloud seeding R&D to meteorology), thus, potentially, ensuring their continuity (and the jobs for the individuals involved), even though they may be transformed in the process.

2.3.2 Case 2: Incandescent Light Bulb

The decline of the incandescent light bulb (ILB) in Europe presents a contrasting narrative to cloud seeding. Before the phase-out of the ILB in Europe, they were very well embedded physically and culturally, which indicates high alignment in the components of the configuration. The European ILB market was mature. In 2006 alone, 2.1 billion out of 5.1 billion bulbs installed in EU households were ILBs (EC 2020). On average, a household had twice as many ILBs as other bulb types (EC 2020). The meaning of the ILB was solidified as a very common object. It was, for instance, the symbol of innovation and bright ideas and, at one point, was popular in jokes of a varying degree of offensiveness.[2] One psychology study found that the light bulb has a traceable psychological and cognitive effect.[3]

At the same time, it also co-existed with other types of lamps. During the 2000s, major manufacturers were selling compact fluorescent lamps (CFL) for niche markets, but were expecting the ILB to continue to dominate at least in the short term (Kelly and Rosenberg 2016). The longer-term expectations were linked with light-emitting diodes (LED). Also, throughout the 2000s, debates around global warming and the role of energy consumption gained steam.

Eventually, a controversy started forming around the ILB, fuelled primarily by national governments and by lamp manufacturers. Led by agendas of meeting climate policy targets and higher profit margins, respectively, these stakeholders managed to transform the meaning of the ILB into that of an energy-inefficient technology that needs to be replaced by more energy-efficient options: first, the CFL, then the LED.

The physical and cultural legacy of the ILB was hard to displace, however, because many light fixtures and lamp users would only accept the ILB shape and the familiar light quality. Thus, new CFL and LED lamps had to imitate the ILB. Teaming up against the ILB, anti-ILB stakeholders pushed the reframing further: is the ILB really

---

[1] A similar point can be found in Van Lente & co-authors: "technologies with generic applications can be linked to a more diverse set of expectations associated with different paths of social embedding" (Van Lente 2013, p. 1617).

[2] "How many [insert name of identity group] does it take to change a light bulb?—Three: one to hold the bulb, and two to turn the ladder" (Kerman 1980).

[3] Elmore and colleagues write that "exposure to an illuminat (ed) light bulb enhances participants' performance on insight-based problems [...] whereas exposure to a burnt-out bulb reduces creativity" (Elmore and Luna-Lucero 2017).

that dear and familiar, or is it rather a "symbol of waste"? Ultimately, the supporters of the ILB were not able to oppose effectively and the ILB sales ban was introduced across the EU. ILBs survive for the most part due to their reframing and continued production and use as "rough-service lamps", e.g. for use in ovens.

Today, regulations, energy-efficiency requirements, and fashion prevent the ILB from scaling up again. The industry and policymakers had weakened the components of the configuration, the hardcore ILB admirers transformed them, supporting the business of the exempt "rough-service" ILBs, while ILB industry factory floor workers bore the highest cost: as the industry shrunk, they had to go through retraining and job-hunting or move to another city or country. Despite niche market presence, there is a clear decline of ILBs on a broader scale, indicative of successful substitution by LEDs and other energy-efficient lighting technologies. This illustrates a successful misalignment of the ILB configuration and decline of the ILB.

In sum, the history of decline of the ILB unfolded as follows:

(a) formation of a negative narrative by a powerful social group and its lobbying efforts to phase out ILBs;
(b) imitation of the ILB by competing LED technology; and
(c) gradual withdrawal of ILBs from mainstream production and use, with exemptions allowing their purchase as "decorative lamps" or "rough-service lamps".

The ILB case shows that there seems to be no need for complete disappearance of the components of the configuration for its decline. In fact, it seems possible for a configuration such as the light bulb to remain in a state of diminished, niche performance for a long time. When discussing the cloud seeding case, I pointed out that some forms of competences seem able to travel to other configurations. The ILB case illustrates that materials can also travel between configurations, leading to e.g. imitation.

2.3.3 Case 3: The Ural Computer

A final case I have studied is the Russian/Soviet original computer series *Ural*. Initially one of the most widely used computers in various sectors of the Soviet economy, like industrial design, medicine, meteorology, and banking during the 1960s, the *Ural* faced a series of critical events that led to its near disappearance by the 1980s. Domestic competition for resources and parts, as well as external pressures in the form of a foreign embargo on parts forced the original design of the *Ural* to be downgraded to inferior materials (punched cards and punched tape). This enabled misalignment in the materials component. Then, the competences component was misaligned after the key members of the *Ural* staff left the project. All this damaged the reputation of the *Ural* and its remaining team and, subsequently, industrial users of the *Ural* withdrew funding. The misalignment of the components created the conditions for the unravelling of the configuration and eventual decline of the *Ural*.

Although some *Urals* continued to be used in specialized sectors like the Soviet space industry into the 1980s, their overall production ceased after the 1970s. The scarcity of resources and space-intensive nature of these computers led to the preservation of only a few of them in museums.

In sum, the history of the Ural's decline may be summarized by the following key events:

(a) departure of the lead designer, which negatively impacted innovation within the *Ural* development team;
(b) government prioritization of a competing series called *ES ÉVM* over other computer projects, indicating a strategic shift in technological investment and support;
(c) resource constraints forcing designers and manufacturers to rely on outdated parts for *Ural* production, which compromised performance and reliability;
(d) hybridization efforts by the design team who attempted to adapt the *Ural* to changing policy demands but potentially undermining its original identity and credibility in the process; and
(e) withdrawal of funding for the *Ural* series.

2.4 Conclusion

Understanding the ageing, dismantling, and phase-out of technologies such as bridges, power plants, and internal combustion engines is sought for in both safety and risk assessment and management and sustainability transitions communities, and a better understanding is needed. In this chapter, I proposed a theoretical framework to conceptualize technological decline, i.e. the process of diminishing production and/or use of a given technology, and backed it up with three illustrative cases. According to this conceptualization, every time a given technology is produced or used, the alignment of three components—materials, collective and individual meanings, and competences—is being reproduced. This is a state of continuity, or dynamic stability of the configuration, and the absence of decline. I proposed that if we conceptualize technological decline as processes within this socio-material configuration, we may shed more light on the mechanisms of decline.

In my cases, I showed the mechanism of misalignment of the components as the breaking apart of the configuration, and the subsequent pathways of either technological decline or re-emergence. With these two pathways, three concepts become important: weak decline, strong decline, and irreversibility of decline. First, categorizing decline into weak and strong may provide a nuanced understanding between the two pathways. Weak decline refers to a situation of relatively low effort and/or cost of re-emergence of technology, where all components are more or less available but are misaligned. Most technologies that colloquially are referred to have "declined", and certainly all three cases discussed above, would fit the category of weak decline.

Strong decline, on the other hand, is more of a hypothetical category which I conceptualized as the disappearance of one or all of the three components. The distinction between weak and strong decline describes irreversibility of decline. The issue of irreversibility is important for such topics as nuclear disarmament, nuclear waste disposal, and sustainability-driven initiatives such as fossil fuels phase-out. Strong decline describes irreversibility, or at least what literature on nuclear disarmament calls functional or "adequate irreversibility" (Ritchie 2023). Since observing—let alone ensuring—absolute or permanent irreversibility is unrealistic, such "adequate irreversibility" refers to a high relative *level* of irreversibility. Because irreversibility almost literally implies a temporal evaluation, I suggest that irreversibility should be analysed with the help of a longitudinal methodology such as the one presented in this chapter.

From the presented framework and case findings, certain practical recommendations may be formulated for the field of risk assessment, safety management and industrial policy, and more broadly towards coordinating or inducing technological decline.

The theoretical framework puts forward an emphasis on the misalignment processes in meanings, materials and competences of the focal configuration. It follows that governance actors interested in coordinating or inducing technological decline (e.g. phase-out) would need to, first, carefully map the given technology as a configuration of meanings, competences and objects and infrastructure, and then put pressure on at least one of the components. For instance, policy makers should have the backing of a coalition of (local) governments, industry and (environmental) NGOs in favour of the phase-out in question. Strong, independent investigative media, as well as a receptive and active parliament, may be key in technological decline, as are trade instruments and international agreements that may be used to leverage other countries' decline of the given technology, as has happened in the case of the *Ural*.

Supporting and/or promoting alternative (existing or emergent) solutions or technologies that fulfil the same need as the technology in question may also be key. This could be done by offering subsidies, tax incentives, direct funding of R&D and (re)training, public relations or lobbying efforts, and compensation to actors who stand to lose from a technology's decline. In fact, a more ethically just decline, albeit at the expense of its irreversibility, is one in which the formerly useful knowledge and skills (and thus the people who are their carriers) are employed in other fields and technologies. An example is the case of the transfer of cloud seeding knowledge to the field of meteorology. Similarly, the case of the light bulb shows how labour unions can cushion the damage caused by decline.

Lastly, actors interested in preventing decline and in preserving the technology in question, should ensure the continued resilience of all three components, e.g. via continued supply of funding, materiel, a skilled and attractively paid workforce, timely physical repair and maintenance, as well as a positive image of the technology in the media.

Naturally, the framework presented in this chapter is just one way to look at the phenomenon that I referred to as "decline" here. The empirical material is for now limited, and the focal points of technology as a configuration and the (mis)alignment of components are necessarily selective. More research on decline and adjacent phenomena and more case studies will surely add on and improve the findings presented here.

References

EC, 'FAQ: phasing out conventional incandescent bulbs'. [Online]. Available: https://ec.europa.eu/commission/presscorner/detail/en/MEMO_09_368. Accessed 16 Oct 2020

K.M. Eisenhardt, M.E. Graebner, Theory building from cases: opportunities and challenges. Acad. Manag. J. **50**(1), 25–32 (2007). https://doi.org/10.5465/AMJ.2007.24160888

K.C. Elmore, M. Luna-Lucero, Light bulbs or seeds? How metaphors for ideas influence judgments about genius. Soc. Psychol. Personal Sci. **8**(2), 200–208 (2017). https://doi.org/10.1177/1948550616667611

M. Foucault, Discipline and punish: the birth of prison. Pantheon Books (1977). https://doi.org/10.2307/2065008

F.W. Geels, Feelings of discontent and the promise of middle range theory for STS: examples from technology dynamics. Sci. Technol. Human Values **32**(6), 627–651 (2007). https://doi.org/10.1177/0162243907303597

F. Goulet, Characterizing alignments in socio-technical transitions. Lessons from agricultural bio-inputs in Brazil. Technol. Soc. **65**, 101580 (2021). https://doi.org/10.1016/j.techsoc.2021.101580

K. Kelly, M. Rosenberg, Some light reading: Understanding trends residential CFL and LED adoption, in *ACEEE Summer Study on Energy Efficiency in Buildings* (2016)

J.B. Kerman, The light-bulb jokes: Americans look at social action processes. J. Am. Folk. **93**(370), 454–458 (1980). https://doi.org/10.2307/539876

Z. Koretsky, Phasing out an embedded technology: Insights from banning the incandescent light bulb in Europe, in *Energy Research and Social Science*, vol. 82, no. 102310 (2021). https://doi.org/10.1016/j.erss.2021.102310

Z. Koretsky, H. Van Lente, Technology phase-out as unravelling of socio-technical configurations: cloud seeding case. Environ. Innov. Soc. Transit **37**, 302–317 (2020). https://doi.org/10.1016/j.eist.2020.10.002

Z. Koretsky, R. Zeiss, H. Van Lente, Exploring the dynamics of technology phase-outs through the history of a Soviet computer "Ural" (1955–1990). Sci. Technol. Hum. Values (2022). https://doi.org/10.1177/01622439221130139

Z. Koretsky, Unravelling: the dynamics of technological decline (Maastricht University, 2022)

Z. Koretsky, Unravelling: The dynamics of technological decline (DataverseNL, 2022). https://doi.org/10.34894/KTHLRV

Z. Koretsky, P. Stegmaier, B. Turnheim, H. Van Lente, (eds.), *Technologies in Decline: Socio-Technical Approaches to Discontinuation and Destabilisation* (Routledge, 2023). https://doi.org/10.4324/9781003213642

B. Latour, Technology is society made durable. Sociol. Rev. **38**(1), 103–131 (1991). https://doi.org/10.1111/j.1467-954x.1990.tb03350.x

T.B. Lawrence, Institutional strategy. J. Manag. **25**(2), 161–187 (1999)

N. Ritchie, Irreversibility and nuclear disarmament: unmaking nuclear weapon complexes. J. Peace Nucl. Disarm. 1–26 (2023)

V. Sheppard, *Research Methods for the Social Sciences: An Introduction* (BC Campus, 2020)

B. K. Sovacool, M. Iskandarova, J. Hall, Industrializing theories: a thematic analysis of conceptual frameworks and typologies for industrial sociotechnical change in a low-carbon future (2023). https://doi.org/10.1016/j.erss.2023.102954

H. Van Lente, C. Spitters, A. Peine, Comparing technological hype cycles: towards a theory. Technol. Forecast Soc. Change **80**(8), 1615–1628 (2013). https://doi.org/10.1016/j.techfore.2012.12.004

Open Access This chapter is licensed under the terms of the Creative Commons Attribution 4.0 International License (http://creativecommons.org/licenses/by/4.0/), which permits use, sharing, adaptation, distribution and reproduction in any medium or format, as long as you give appropriate credit to the original author(s) and the source, provide a link to the Creative Commons license and indicate if changes were made.

The images or other third party material in this chapter are included in the chapter's Creative Commons license, unless indicated otherwise in a credit line to the material. If material is not included in the chapter's Creative Commons license and your intended use is not permitted by statutory regulation or exceeds the permitted use, you will need to obtain permission directly from the copyright holder.

Chapter 3
Discontinuation Through Enforcement of the Law: Court Rulings as Leverage for Stopping Delegitimized Practices and Technologies

Peter Stegmaier

Abstract Not everything that is done is legal. What is not legal but is done anyway is often ignored or seen as grey areas. With the help of court decisions, a strict interpretation of the law leads to existing practices being transferred from being ignored to being relevant, with the result that existing (illegal?) institutional agreements and practices are no longer tolerated. Processes can be observed in which the combined resources of law, legal representation, and courts are used to promote steps towards discontinuation of environmentally harmful practices and infrastructures. These involve the mobilization of non-political means and ways to make governance and policy. The cases help to explore very different constellations and contexts, all with side-consequences that might lead to discontinuation. Before ageing systems can be shut down and declining technologies discontinued, their existence, maintenance, and use "as usual" must first be delegitimized. One way in which such work on discontinuation can proceed is presented here.

Keywords Discontinuation governance · Law · Delegitimization

3.1 Introduction: Forcing Issues from Ignorance to Relevance

Climate litigation initiatives have been emerging in many countries around the world for some years. The 2023 UN Global Climate Litigation report states that "climate change litigation is increasing and broadening in geographical reach, while the range of legal theories is expanding" (UNEP 2023: XIV). Not all climate lawsuits and not in all respects are aimed at actions that should be stopped. It is also about encouraging new action, actively pursuing transitions, and taking concrete measures. Where this

P. Stegmaier (✉)
Knowledge, Transformation and Society Research Group, University of Twente, Enschede, The Netherlands
e-mail: p.stegmaier@utwente.nl

© The Author(s) 2025
M. Bourrier (ed.), *Decommissioning Aging Installations and Declining Technologies*, SpringerBriefs in Safety Management, https://doi.org/10.1007/978-3-031-88369-9_3

is not considered sufficient to reduce environmentally damaging behaviour, rights activists in particular insist that such behaviour be stopped. Here we have cases in which *the (new) interpretation of the law transfers existing practices from ignorance to relevance—with the consequence of existing (illegal?) institutional arrangements and practices no longer being tolerated.* The strategy can be observed that the climate problem is used as a discursive lever to take action. Processes can be observed in which the combined resources of law, legal representation, and courts are used to promote steps towards discontinuation of environmentally harmful practices (e.g. air pollution from road transport, energy production) and infrastructures (e.g. roads used by polluting vehicles, exhaust systems, extraction, and energy production facilities and networks). These processes mobilize non-political means and ways to make governance and policy. This is where we want to get to the bottom of discontinuation practices: discontinuation work using law.

This topic is part of my research on active discontinuation of socio-technical systems and their regimes. In this ongoing project, I seek out the most diverse practices and practical contexts in which "doing discontinuation" can be observed and understood as *governance-in-action* (Stegmaier et al. 2014). Ideally, this complements the research of other colleagues and me from STS (Koretsky 2023), transition studies (Turnheim 2023), and more systemic perspectives (Turnheim and Geels 2012; Isoaho and Markard 2020). I want to provide a perspective to look beyond the relatively specific field of legal litigation (Benjamin 2021; Ganguly et al. 2018; Meltz 2007) to the environmental circumstances and strategies and how these are even becoming established as a professional competency.

In this chapter, I compare the inner-city ban on diesel vehicles in North Rhine-Westphalia, Germany (February 2018), the strengthening of the Climate Change Act through the Federal Constitutional Court in Germany (March 2021), and the decision in the case of Milieudefensie et al. v. Royal Dutch Shell plc. in the Netherlands (May 2021). The cases serve to explore very different constellations and contexts. All the decisions have indirect consequences that might lead to discontinuation. The first case of diesel driving bans in particular also led to direct bans and the prohibition of diesel car use on certain roads. At the end of the chapter, I make references to another practical area of discontinuation work, dismantling. For both areas, I would like to suggest interpreting the developments in such a way that discontinuation emerges as professional behaviour.

3.2 The Perspective of Studies on Discontinuation Governance-Making

Studies of governance-in-action aimed at the discontinuation of socio-technical regimes and niche developments (practices, institutional arrangements, policies, infrastructures, investments, etc.) have so far mainly focused on (1) governance processes and strategies for negotiating problem foci and agendas, policy alternatives and policy choices, time frames, actor constellations, legitimations (Stegmaier 2023; Stegmaier et al. 2021), (2) the negotiation of discontinuation in formal and informal arenas (from cabinet work to street blockade, from protest back to policy, from negotiation to interdiction; (*cf.* Liersch and Stegmaier 2022; Stegmaier et al. 2014), (3) the development of new regulation to implement discontinuation goals, and (4) destabilization as an accumulation of circumstances leading to the disintegration of existing regimes (Koretsky et al. 2023). In this chapter, I shift the focus to court proceedings to mobilize existing laws, literally forcing discontinuation through the courts, what I call *discontinuation through judicial action*.

I understand discontinuation to mean a property of a trajectory in which the constituting relations become misaligned to such an extent that its distinctive character is lost, as one possible result of various permutations of distributed agency, contingency, emergence, or deliberate governance (*cf.* Stegmaier 2023). Discontinuation governance involves two complementary aspects, in which groups of governance actors (as insiders and outsiders) undertake towards and ultimately within a window of opportunity a repertoire of actions (including use of parallel policy instruments) intended to affect both the discontinuation of a trajectory itself (governance of discontinuation) and of governance practices that help stabilize it (discontinuation of governance).

3.3 The Cases in Comparison

The three selected cases help to give an impression of *actively pursued discontinuation governance*, which uses the leverage of court decisions. In all three cases, these court judgements constitute decisive turning points. Looking at transformation and focusing on the particular track of change towards discontinuation, these are decisive moments when a breakthrough is achieved, a critical mass of events or coalition partners come together, or circumstances change significantly. From an active governance-making perspective, these can be described as opportunities that occur, are recognized, and seized (or not) in limited time windows or phases.

3.3.1 *Inner-City Bans on Diesel Vehicles*

Main focus

Our first case, the ban on diesel vehicles on city centre roads, is interpreted as a step against the tacit acceptance of illegal emissions from diesel engines in cars. Applicable law is first ignored, and then compliance is enforced: Courts confirm that the law must be complied with, and that politics and administrations must ensure compliance across the board, i.e. for the entire urban areas in which emission limits are exceeded—with immediate and permanent measures.

Main characteristics

Limits on nitrogen oxide concentrations in the air are exceeded in numerous cities in North Rhine-Westphalia, Germany, for instance. The Environmental Action Germany (DUH) initiative took this as an opportunity to sue the state of North Rhine-Westphalia. The Gelsenkirchen Administrative Court has ruled that from 1 July 2019, diesel cars, buses, and commercial vehicles up to and including the Euro 4 emissions standard were banned from driving on the A40 motorway, extended to Euro 5 diesel from 1 September 2019 onward. It justified the decision, among other things, by the high level of pollution in a residential area in Essen-Frohnhausen, which is located alongside the motorway. The court also ordered the state to examine further driving bans for nine other suspected cases outside the existing "environmental zone" with a deadline of 1 April 2019. This is not the first such lawsuit that DUH has won.

Main consequences

Elsewhere, state governments had appealed against lower instance court decisions and administrative measures taken by cities, but these appeals were not upheld by Germany's highest administrative court. The Federal Administrative Court obliged the countries of North Rhine-Westphalia & Baden-Württemberg in February 2018 to enable the necessary measures to be taken to comply as quickly as possible with the limit value for nitrogen dioxide (NO_2) of 40 $\mu g/m^3$ in urban areas, averaged over one year, by amending the clean air plan for Düsseldorf and Stuttgart on the basis of the Clean Air Plan, the Federal Emission Control Act (BImSchG), and EU regulations 96/62/EG, 2008/50/EG, 1999/30/EG. The claimant in both cases was again DUH. As a result, numerous cities and federal states found themselves obliged to impose driving bans.

3.3.2 Strengthened Climate Protection Act

Main focus

Our second case can be interpreted as a step against climate protection legislation that does not provide for sufficient emission reductions from 2031. In 2020, four constitutional complaints were filed by a group of individuals, supported by various environmental NGOs,[1] against the Federal Climate Protection Act (KSG). The defendants were the parliament and the federal government of Germany. In March 2021, the Federal Constitutional Court ruled that the provisions of the Federal Climate Protection Act (KSG) of 12 December 2019 on national climate targets and annual

---

[1] The group of plaintiffs consisted of the German Solar Energy Association (Solarenergie-Förderverein Deutschland, SFV), the German Federation for the Environment and Nature Conservation (Bund für Umwelt und Naturschutz Deutschland, BUND) and the individual plaintiffs of the lawsuit filed in 2018. This was followed by other individuals and associations with their own lawsuits in 2020. 15 people from Bangladesh and Nepal were also recognized as having the right to lodge a complaint.

emission levels permitted until 2030 are incompatible with fundamental rights insofar as they do not contain sufficient requirements for further emission reductions from 2031.

The German Climate Protection Act provides for a 55 per cent reduction in national carbon dioxide emissions by 2030. Following a complaint, the Federal Constitutional Court obliged the legislator for the first time in the country's legal history to take timely precautions, including for the period after 2030, in order to achieve the transition to climate neutrality in 2050.

Main characteristics

The court refers to Article 20a of the German Basic Law, which states that the state must protect the natural foundations of life, including responsibility for future generations, and explains that.

> Article 20a is a justiciable legal provision designed to commit the political process to a favouring of ecological interests, partially with a view to future generations who will be particularly affected. (BVG 2021a: 3)[2]

The article is thus a sharp legal sword, a provision that can be used to sue for intergenerational justice in climate protection.

The court is not calling for an immediate radical reduction in greenhouse gas emissions, but for the protection of civil liberties for all those who will be restricted in their civil liberties by later restrictions. So, it is not about the restrictions caused by climate change itself, but by the later unavoidably considerably stricter state climate protection measures if too little climate protection is done now. This means that the judgement emphasizes the framework set by the climate protection requirement in Article 20a and calls for the climate protection law to specify requirements that allow the objective to be met. On the other hand, it indirectly suggests that climate-damaging laws and consequences of laws, behaviours and policies must be ended over a reasonable period of time that preserves the freedom of future generations.

Main consequences

What has now been discontinued? Firstly, a law was found to be unconstitutional in parts and its amendment was demanded. This means that it may no longer apply beyond a certain cut-off date. Indirect discontinuation consequences are more in the nature of meta-governance: the court demands "development pressure" for climate-neutral solutions and "planning security" in order to be able to initiate the transition to climate neutrality "in good time". Measures of all kinds are therefore required that enable climate-friendly behaviour and stop climate-damaging behaviour. The latter is the area in which the constitutional court ruling can be expected indirectly because CO_2 emissions, technologies, policies, and business models must be banned more or less directly in order to achieve the goals of a climate protection law. Time will tell what levers actors will find to either prevent such practices in future or to water down the far-reaching consequences of the decision in some areas.

---

[2] "Art. 20a GG ist eine justiziable Rechtsnorm, die den politischen Prozess zugunsten ökologischer Belange auch mit Blick auf die künftigen Generationen binden soll". (BVG 2021b: 3).

The model for this case, treated internationally also as Neubauer, et al. v. Germany, could be the action brought by the Urgenda Foundation seeking declaratory judgement and injunction to compel the Dutch government to reduce greenhouse gas emissions, also known as Urgenda Foundation v. State of the Netherlands.[3]

3.3.3 Milieudefensie et al. V. Royal Dutch Shell Plc.

Main focus

Our third case can be interpreted as a step against business behaviour that puts Dutch residents at risk from aggravated climate change, or as an observer has put it, as an answer to the legal-societal question: "Whether a private company violated a duty of care and human rights obligations by failing to take adequate action to curb contributions to climate change" (Climate Change Litigation Databases 2023). In 2019, Milieudefensie, the Dutch branch of Friends of the Earth International, alongside six other organizations and 17,000 Dutch citizens took legal action against Royal Dutch Shell plc. based on Dutch tort law, civil law, and international law. In May 2021, the District Court of the Hague (Rechtbank Den Haag 2021) decided that the Shell Group is responsible not only for its own CO_2 emissions but also those of its suppliers and clients. It also decided that Shell must reduce these CO_2 emissions by 45% by 2030 compared to 2019 (NRC 2021a; BBC 2021; Climate Change Litigation Databases 2023). The court found that.

> plaintiffs could bring a class action in the common interest of preventing dangerous climate change, since the interests of current and future generations of Dutch residents 'are suitable for bundling so as to safeguard an efficient and effective legal protection of the stakeholders' …. (Nollkämper 2021)

Shell appealed the decision on 20 July 2022.

Main characteristics

The special features of the judgement are that it is the first of its kind in the Netherlands (innovation in case law; *cf.* Mayer 2022, 2021), that the judge holds a company to agreements that were not made by the company, but by governments, that climate protection is judicially interpreted as a human rights issue (climate protection is given high symbolic legitimation), and that the decision made in Den Haag has a global juridical scope (*cf.* Spijkers 2021). Finally, it is an environmental organization of citizens that has triggered the judgement against the company (informal politics through legal-formal channels; *cf.* Hösli 2021).

---

[3] Cf. https://climatecasechart.com/non-us-case/urgenda-foundation-v-kingdom-of-the-netherlands/; www.urgenda.nl/en/themas/climate-case/

Main consequences

Although the company does not state it as a consequence of the judgement, Shell dropped the name "Royal Dutch" from its name after the defeat in the District Court of The Hague and moved its headquarters to London, UK (NRC 2021b; FT 2021). In November 2024, the Court of Appeal followed the lower court and ruled that Shell does have a legal duty of care to mitigate dangerous climate change under Dutch tort law in conjunction with international human rights instruments and EU and international climate law, but overturned the ruling by refusing to impose a specific emissions reduction target on Shell. It justified this on the grounds that it did not see sufficient scientific consensus on a specific reduction percentage or path that a single company such as Shell should adhere to.

3.3.4 Main Lines of Comparison

First line of comparison: discontinuation targets

In comparing the cases, we have car emissions, national emissions, and corporate emissions as targets of legal action. This makes a whole range of actors responsible, mobilizes very different areas of law and varies the depth of impact. The latter means that the driving bans will have a very direct effect (promptly, at specific locations), while the verdict against excessively lax environmental protection legislation will have a more indirect effect, over longer periods of time and mediated via more specific measures that still need to be taken.

Court rulings are often permanently binding but can be revised from higher instances or developed in other directions on the basis of new facts (cases) even by lower instances. Law is relatively stable, but not timeless, not independent of the socio-historical point of view. If one therefore considers the negotiating character of judgments, the objects of dispute and judgments appear to be *boundary objects of a long-term development*: of (a) public emissions monitoring and limits and their formal objectivations, (b) a public law and its violation of fundamental rights, and (c) corporate emissions (and their formal objectivations). This development still needs to be further investigated empirically.

Second line of comparison: role of the law and the court

In the case of the diesel driving ban, the existing emissions protection and air pollution control law was forced to be applied in specific areas of emissions pollution with a very specific technology as the polluter. The court was able to apply existing law and case law in the usual way. In the case of the strengthening of the accuracy of the Climate Protection Act, the modification of a law was enforced by clarifying the meaning of a fundamental constitutional norm ("freedom"). The core constitutional concept was redefined, as Ulrich (2021: 3) explains:

> Freedom is now no longer just something that you live and can assert against the state, freedom is now also something that you can consume materially and physically. And what you must not do.[4]

The court has replaced or at least supplemented the "ecologically blind concept of freedom" (ibid.) with an ecologically anticipatory, responsible concept of freedom. In the third case, the connection between human rights and climate protection has been emphasized, as is the recently reinforced compliance with human rights standards in industrial production, particularly outside the EU (Directive on corporate sustainability due diligence through Directive 2024/1760).

Application of the law and further development of the law are the reference points of the three judgments, which were realized in different ways: In the case of the diesel driving bans, the law was enforced by taking measurements and drawing the prescribed consequences. In the case of the Climate Protection Act, the fundamental legal justification was worked out in an innovative way. In the case of the lawsuit against Shell, human rights and climate protection were linked in accordance with the polluter pays principle.

3.4 Discussion

In the following, I would like to address some further aspects that were not or only indirectly alluded to in the direct comparison.

Firstly, previous case studies have shown that discontinuation governance measures can be part of a chain of escalating steps of an increasingly broad, sustainable, and fundamental nature. In very simplified terms, this can be visualized as a *spiral of discontinuation.* Discontinuation tends to begin with control measures through systematic *monitoring*, the creation of knowledge about an area of a sociotechnical system, which can be used to justify further measures if necessary. This is followed by the *restriction* of the scope of usage (if the expectation that monitoring will take place did not already lead to self-restriction), and later the limitation of the scope of production (discontinuation by *reduction*). The cascade is often completed by a combination of immediate *bans* on sub-areas that are considered particularly harmful or easy to implement as well as the gradual *phasing-out* of other sub-areas that are considered less urgent or more difficult to implement (unless discontinuation cannot be enforced successfully or further intermediate steps are introduced). These processes are usually not linear, but tend to intensify in several rounds and in different sequences.

The judicial measures under consideration have an impact on different depths of governance intervention when the spiral of discontinuation is set in motion. The diesel ban on inner-city roads is similar to the restriction of the use of a technology in the

---

[4] The original quote: "Freiheit ist nun nicht mehr nur etwas, das man lebt und gegen den Staat geltend machen kann, Freiheit ist nun auch etwas, das man materiell und physisch verbrauchen kann. Und nicht darf". (Ulrich 2021: 3).

area of consumer products, which takes place before the restriction of production. The ban on diesel vehicles is a limited cessation of a technology whose use in a specific environment is no longer considered legal. Not diesel driving as such, but on certain roads; not driving with internal combustion engines in general, but specifically with diesel, because collectively this causes certain emissions, the level of which regularly exceeds a legally defined limit.

Secondly, for all players in politics and business, these cases show that there are endeavours towards discontinuation that do not simply emerge from a given political field out of the blue but have solid legal foundations. Those who want to foresee future areas and steps of discontinuation have to follow the currently emerging constellations in the fields of interest as well as have a good overview of the existing institutional framework conditions and anticipate mutually beneficial relationships—design and play through *discontinuation scenarios.*

Companies and political rivals in particular indeed do this, and so it is not surprising that, conversely, one could study in many cases how concerted actions via political lobby channels and media forums regularly aim to discredit and otherwise undermine nascent or already flourishing discontinuation proposals or measures—one could call this *anti-discontinuation policy.*

Thirdly, business, governments, corporations, and other organizations that can be addressed as legal entities can *expect to be the target of legal action* if they act against the interests of climate protection. NGOs have been successful several times in numerous countries with lawsuits against climate polluters. They are now identifying new cases and preparing actions. If organizations or public bodies are engaged in climate-damaging activities, it would be naïve for them to assume that they will not sooner or later become the target of active resistance through legal and judicial means. I already mentioned MOB in the Netherlands now targeting provinces and livestock farmers and their nitrogen permits (NOS 2023). Dutch Milieudefensie, after Shell, is now targeting ING, the largest Dutch bank (NRC 2024): the action group demands a climate plan from ING and associated concrete measures. These measures mean the bank would have to halve its CO_2 emissions by 2030 compared to 2019. According to Milieudefensie, ING's current policy is undermining the achievement of the maximum of 1.5 degrees of global warming, as agreed in the Paris climate agreement. Milieudefensie argues that "ING has the largest CO_2 footprint of all financial institutions in the Netherlands. Its footprint is equal to that of all companies and households in Sweden combined" (ibid., *own translation*). Milieudefensie explains its strategy as combating global warming where it has the most influence, namely in large companies. In 2024 they have published a target list of the 29 biggest polluters in the Netherlands.

Environmental Action Germany (DUH) uses the language of an executive body when it defines its work as "We enforce nature and environmental protection".[5] DUH is an environmental and nature conservation association and at the same time a recognized consumer protection association with the right to sue, which is on the list of qualified organizations at the German Federal Office of Justice. DUH is also

---

[5] www.duh.de/ueberuns/; www.duh.de/englisch/who-we-are/

a recognized nature conservation organization for actions before the administrative courts. This means that its *litigation work is authorized by law* in the areas of environmental, nature and consumer protection. A large proportion of DUH's work is associated with campaigning for more protection and against pollution. In both cases, legal action is taken to put an end to harmful behaviour (discontinuation work) and/or to enforce protective measures (behaviour change work).

Discontinuation through court decisions could be seen as the expression of a *normalization of discontinuation as a practice*. More cases and contexts need to be studied. It will also be interesting to see how these cases stand not only for discontinuation of an environmentally harmful practice itself (e.g. driving through cities with air-polluting vehicles), but also of framing practices that stabilize it (e.g. deliberately not noticing illegal levels of exhaust emissions, speed beyond the limit, etc.).

3.5 Conclusions and Outlook

The exploration of the three cases has shown that there is organised activity aimed at putting an end to environmentally harmful behaviour. Organizational actors such as Milieudefensie, MOB, and DUH in the Netherlands and Germany have developed fields and forms of work that are essentially aimed at using the rule of law, law enforcement, and legal action to stop practices and technology. This goes far beyond the mobilization of the public, agenda setting, protest organization, etc.

So far in this chapter, we have looked specifically at the legal work of enforcing discontinuation, which is becoming increasingly professionalized, financed, and organized. In the outlook, I would like to highlight the wider spectrum of activities that should not be overlooked in this context: some other forms of organized discontinuation work. From this I derive the thesis that legal discontinuation work is an expression of an institutionalizing orientation of practice: dismantling as job, business, vocation, and emerging social field.

There is a vigilant range of study programmes and the development of job profiles that are linked to the dismantling of nuclear power plants, to address a completely different area of discontinuation work. At the Czech Technical University in Prague, specialists for the decommissioning of nuclear power plants are trained in the master's programme "Decommissioning of Nuclear Facilities" since 2022. The Technical University Dresden has launched the graduate programme "Sustainable Decommissioning of Power Plants—Challenges and Solutions for Technology, Environment and Society". The EU Academy offers a course on the "Safe Decommissioning of Nuclear Facilities". Discontinuation can now be studied. At the Karlsruhe Institute of Technology (KIT) a Competence Center for Decommissioning was established in 2015. One can also find jobs in the field of business management and asset management, which—not as an entire job profile, but as an area of responsibility—deal with the handling of exits from investments, corporate divisions, shareholdings, etc.

In most cases (including those referred to above), there are job descriptions that exclusively determine *discontinuation as a personal identity*, for instance, of liquidators, such as "construction mechanic for demolition and concrete cutting technology" or "site manager for dismantling/demolition work and refurbishment". The aforementioned site manager job also includes refurbishment, while the demolition mechanic rather seems to focus on dismantling only. In various countries, there are industry associations with a focus on demolition. In Germany, where a lot of demolition work is currently being carried out on nuclear power plants, this is the German Demolition Association (DA),[6] which has been in existence since 1951 and represents the interests of businesses, entrepreneurs, and employers in the demolition industry in Germany and throughout Europe. It advertises and offers vocational trainings for careers in the demolition industry.[7] Whether and to what extent a kind of discontinuation profile or even an industry of "liquidators or terminators of delegitimized socio-technical and socio-economic systems" will emerge as a governance-making profession or legal profession remains to be seen empirically.

I would like to call this orientation of activities *discontinuation work*: the structured work in an organized context on the stopping, decomposition, or termination of an object described as a goal (behaviour, business model, investment, governance arrangement, socio-technical system, technological device of hard-/soft-/socioware). I am referring primarily to the work activity and action with the declared aim of leaving an impact. I consider it primarily insightful to look at *how discontinuation is carried out*.

Whether the discontinuation of socio-technical systems that are considered obsolete, harmful, or no longer functional or desirable represents a burden or a positive impetus is, from a social science perspective, initially a *question of perspective*: of the observer, the affected, the profiteers and losers, the informed, the uninformed, the misinformed, the dependent, and the independent. In practice, exits can probably not be achieved without effort, undesirable side effects, complications, and the like. They can only be planned to a limited extent. Civil society, politics, administration, management, and the media must take this into account. But they can also build on growing professional expertise.

References

BBC, (2021), *Shell: Netherlands court orders oil giant to cut emissions* (26 May 2021). www.bbc.com/news/world-europe-57257982. (22 Aug 2024)

L. Benjamin, *Companies and climate change: Theory and law in the United Kingdom.* (Cambridge University Press, 2021)

---

[6] www.deutscher-abbruchverband.de/en/verband/verbandsportrait-und-geschichte/ [22 August 2024].

[7] www.deutscher-abbruchverband.de/en/career/education-and-continuing-education/ [22 August 2024].

BVG, (2021a), Constitutional complaints against the Federal Climate Change Actpartially successful. Press Release No. 31/2021 of 29 April 2021. Karlsruhe: BVG. www.bundesverfassungsgericht.de/SharedDocs/Pressemitteilungen/EN/2021/bvg21-031.html. (22 Aug 2024)

BVG, (2021b), Verfassungsbeschwerden gegen das Klimaschutzgesetz teilweise erfolgreich. Pressemitteilung Nr. 31/2021 vom 29. April 2021. www.bundesverfassungsgericht.de/SharedDocs/Pressemitteilungen/DE/2021/bvg21-031.html. (22 Aug 2024)

C. Higham, J. Setzer, E. Bradeen, (2022), *Challenging government responses to climate change through framework litigation.* London: Grantham Research Institute on Climate Change and the Environment and Centre for Climate Change Economics and Policy, London School of Economics and Political Science. www.lse.ac.uk/granthaminstitute/publication/challenging-government-responses-to-climate-change-through-framework-litigation/. (22 Aug 2024)

Climate Change Litigation Databases, (2023), *Milieudefensie et al. v. Royal Dutch Shell plc.* https://climatecasechart.com/non-us-case/milieudefensie-et-al-v-royal-dutch-shell-plc/. (22 Aug 2024)

FT, (2021), Shell investors back moving HQ from Netherlands to UK, 10 December 2021, in Financial Times. At: www.ft.com/content/d932a462-2b31-479a-bd08-6e8abc02c375. (22 Aug 2024)

G. Ganguly, J. Setzer, V. Heyvaert, If at first you don't succeed: suing corporations for climate change. Oxf. J. Leg. Stud. **38**, 841–868 (2018)

R. Den Haag, (2021), C/09/571932/HA ZA 19–379 (Engelse versie), 26 May 2021. https://linkeddata.overheid.nl/document/ECLI:NL:RBDHA:2021:5339. (22 Aug 2024)

A. Hösli, Milieudefensie et al. v. Shell: a tipping point in climate change litigation against corporations? Climate Law **11**:195–209 (2021)

K. Isoaho, J. Markard, The politics of technology decline: discursive struggles over coal phase-out in the UK. Rev. Policy Res. **37**, 342–368 (2020)

Z. Koretsky, Dynamics of technological decline as socio-material unravelling, in *Technologies in Decline: Socio-Technical Approaches to Discontinuation and Destabilisation.* ed. by Z. Koretsky, B. Turnheim, P. Stegmaier, H. van Lente (Routledge, London, 2023), pp.15–42

Z. Koretsky, P. Stegmaier, B. Turnheim, H van Lente (ed.), *Technologies in Decline: Socio-Technical Approaches to Discontinuation and Destabilisation* (Routledge, 2023)

C.C. Liersch, P. Stegmaier, Keeping the forest above to phase out the coal below: the discursive politics and contested meaning of the Hambach Forest. Energy Res. Soc. Sci. **89**, 1–19 (2022)

B. Mayer, The duty of care of fossil-fuel producers for climate change mitigation. milieudefensie v. royal dutch shell district court of The Hague (The Netherlands). Transnatl Environ Law **11**, 407–418 (2021)

B. Mayer, Judicial interpretation of tort Law in Milieudefensie v. Shell: a Rejoinder. Transnatl Environ Law **11**, 433–436 (2022)

R. Meltz, Climate Change Litigation: A Growing Phenomenon. CRS Report for Congress. (Congress Research Service, 2007)

A. Nollkämper, (2021), Shell's responsibility for climate change. An international law perspective on a Groundbreaking judgment, in Verfassungsblog.de. At: https://verfassungsblog.de/shells-responsibility-for-climate-change/. (22 August 2024)

NOS, (2023), Milieugroep MOB zet provincies onder druk om vergunningen veehouders in te trekken, in NOS, 28 March 2023. https://nos.nl/artikel/2469248-milieugroep-mob-zet-provincies-onder-druk-om-vergunningen-veehouders-in-te-trekken. (22 Aug 2024)

NRC, (2021a), Deze uitspraak over Shell laat zien: het klimaat is een mensenrechtenkwestie, in *NRC*, 26 May 2021. At: www.nrc.nl/nieuws/2021/05/26/uitspraak-over-shell-laat-zien-het-klimaat-is-mensenrechtenkwestie-a4045058. (22 Aug 2024)

NRC, (2021b), Shell vertrekt definitief naar Londen: aandeelhouders stemmen voor verhuizing, in *NRC*, 10 December 2021. At: www.nrc.nl/nieuws/2021/12/10/shell-vertrekt-definitief-naar-londen-aandeelhouders-stemmen-voor-verhuizing-a4068536. (22 Aug 2024)

NRC, 2021c, Olieconcerns staan nu van alle kanten onder druk, in *NRC*, 21 May 2021. At: www.nrc.nl/nieuws/2021/05/21/olieconcerns-staan-nu-van-alle-kanten-onder-druk-a4044504. (22 Aug 2024)

NRC, (2024), Milieudefensie pakt na Shell nu ING aan via de rechter: 'Stoppen met olie in 2040? Dat is vijftien jaar te laat', in *NRC*, 19 January 2024. At: www.nrc.nl/nieuws/2024/01/19/milieudefensie-pakt-na-shell-nu-ing-aan-via-de-rechter-stoppen-met-olie-in-2040-dat-is-vijftien-jaar-te-laat-a4187529. (22 Aug 2024)

de. Rechtpraak, (2021), Royal Dutch Shell must reduce CO2 emissions (26 May 2021), in *de Rechtpraak*. www.rechtspraak.nl/Organisatie-en-contact/Organisatie/Rechtbanken/Rechtbank-Den-Haag/Nieuws/Paginas/Royal-Dutch-Shell-must-reduce-CO2-emissions.aspx#:~:text=The%20Hague%20District%20Court%20has,the%20Shell%20group%27s%20corporate%20policy. (22 Aug 2024)

M. Redelfs, Die Gegner der Energiewende, (2021), in *Greenpeace Investigative Recherche*, 12 February 2021. www.greenpeace.de/ueber-uns/leitbild/investigative-recherche/netz-windkraftgegner. (22 Aug 2024)

O. Spijkers, Friends of the Earth Netherlands (Milieudefensie) v Royal Dutch Shell. Chin. J. Environ. Law **5**, 237–256 (2021)

P. Stegmaier, Conceptual aspects of discontinuation governance: an exploration, in *Technologies in Decline: Socio-Technical Approaches to Discontinuation and Destabilisation*. ed. by Z. Koretsky, B. Turnheim, P. Stegmaier, H. van Lente (Routledge, London, 2023), pp.78–115

P. Stegmaier, S. Kuhlmann, V.R. Visser, The Discontinuation of Socio-Technical Systems as Governance Problem, in *Governance of Systems Change*. ed. by S. Borrás, J. Edler (Elgar, Cheltenham, 2014), pp.111–131

P. Stegmaier, V.R. Visser, S. Kuhlmann, The incandescent light bulb phase-out. Exploring patterns of framing the governance of discontinuing a socio-technical regime. Energy, Sustain. Soc. **11**, 1–22 (2021)

B. Turnheim, Destabilisation, decline and phase-out in transitions research, in *Technologies in Decline: Socio-Technical Approaches to Discontinuation and Destabilisation*. ed. by Z. Koretsky, B. Turnheim, P. Stegmaier, H. van Lente (Routledge, London, 2023), pp.43–77

B. Turnheim, F.W. Geels, Regime destabilisation as the flipside of energy transitions: lessons from the history of the British coal industry (1913–1997). Energy Policy **50**, 35–49 (2012)

B. Ulrich, (2021), Die Befreiung der Freiheit. Die ZEIT, 30 April 2021, www.zeit.de/politik/deutschland/2021-04/karlsruhe-bundesverfassungsgericht-klimaschutz-urteil-grundgesetz-freiheit/komplettansicht (22 Aug 2024)

Open Access This chapter is licensed under the terms of the Creative Commons Attribution 4.0 International License (http://creativecommons.org/licenses/by/4.0/), which permits use, sharing, adaptation, distribution and reproduction in any medium or format, as long as you give appropriate credit to the original author(s) and the source, provide a link to the Creative Commons license and indicate if changes were made.

The images or other third party material in this chapter are included in the chapter's Creative Commons license, unless indicated otherwise in a credit line to the material. If material is not included in the chapter's Creative Commons license and your intended use is not permitted by statutory regulation or exceeds the permitted use, you will need to obtain permission directly from the copyright holder.

Chapter 4
Undesirable Systems, Undesired Ends: The (Un)Bearable Heaviness of Phasing-Out Pesticides?

Bruno Turnheim, Marc Barbier, and Mireille Matt

Abstract Abandoning harmful technologies and associated practices generating various types of risks is a considerable challenge. It supposes to address the discontinuation of socio-technical systems as a new class of governance problem. The phase-out of large socio-technical systems calls for the joint consideration of systemic structuration and the temporalities of socio-technical transformations. Phasing-out of a toxic world conjures up an overlay of different forms of risk, which requires new takes on the scales, temporalities, politics and uncertainties involved, as well as new conceptions of risks and how they can be handled. Focusing specifically on the removal of pesticides use from agriculture in France (1945–present), this chapter seeks to understand how injunctions to emancipate from dependence to harmful systems have been processed within society. Understanding the successive attempts to ban, reduce or phase-out pesticides request to appreciate the emergence of framings and their translations in a longitudinal analysis. We propose to consider the socio-historical conditions that can explain their (un)felicitous development in three phases of designating and problematizing pesticides: (1) as dangerous substance, (2) as object of reduction and removal, (3) as socio-political risk. We propose a retrospective exploration of the conditions of unveiling pesticide uses at a systemic level to issue the existence of a socio-political risk of not fulling the promise of opening a new era for agri-food systems. Phasing-out pesticides, like dealing with other kinds of persistent systemic issues, appears as if imbued with a still currently unbearable heaviness. Confronted with what appears as the impossible task of phasing-out undesirable (risky) systems while ensuring that such (risky) ends are not undesired, it becomes clear that what is needed is an appropriate framing of the new kinds of socio-political risks that are associated with effective phase-out governance.

B. Turnheim (✉) · M. Barbier · M. Matt
Laboratoire Interdisciplinaire Sciences Innovations Sociétés (UMR LISIS), INRAE-UGE-CNRS, Marne-La-Vallée, France
e-mail: bruno.turnheim@inrae.fr

B. Turnheim
Manchester Institute of Innovation Research (MIoIR), University of Manchester, Manchester, UK

Keywords Transitions · Discontinuation of socio-technical systems · Governance of systemic transformations · Longitudinal analysis · Phase-out and reduction of pesticides · Socio-political risks of transition

4.1 Introduction

Abandoning harmful technologies, associated practices or underlying socio-technical systems is a considerable challenge of our time, one that bears relevance to a variety of sectors, including energy, food, mobility or industrial processes (Koretsky et al. 2023). In contrast to a Schumpeterian approach of innovation as a process of creative destruction, the harrowing hangover that the Anthropocene presents us with opens radically different prospects for the discontinuation of socio-technical systems, since it calls for an intentional governance of systemic transformations. From such a standpoint, we can appreciate how systems framed as harmful (or "undesirable") are becoming the object of phase-out politics, policies and strategies.

The discontinuation of socio-technical systems—which includes system phase-out—points to a class of governance problems (Stegmaier et al. 2014) that significantly differ from the mitigation of risks posed by human activities (and in particular large-scale productive activities) or from the development and deployment of (less harmful) innovation. Indeed, according to such a perspective, the persistent forms of risks associated with certain industries and their products (e.g. fossil fuels, nuclear power, pesticides, etc.) cannot be dealt with effectively through incremental environmental improvements or substitutes and instead call for dedicated strategies. Phase-out has become an object of research, particularly within the literature on transitions studies, which harbours a dedicated stream of research on destabilization, decline and phase-out (Turnheim 2023a).

The starting proposition of this chapter is that abandonment, reduction and removal differ significantly from adoption, expansion and addition at play in innovation processes at the heart of socio-technical regime. Since phase-out appears necessary but complex, then putting it into action is not at all trivial. Phasing-out any socio-technical system calls then for dedicated heuristics and dedicated governance strategies, that deserve to be studied and understood systematically in their socio-historical contexts, as has been done elsewhere for the dismantling of tramways (Turnheim 2023b), or as we are going to present it in this chapter dedicated to the phase-out and reduction of pesticides.

While focusing specifically on the removal of pesticide use from agriculture in France, this chapter seeks to understand how injunctions to emancipate from dependence to harmful systems have been processed within society throughout and beyond the "green revolution". Before contributing to the research agenda on pesticides (Mansfield et al. 2024) with some transition studies insights, let's first present some theoretical background to clarify how to approach phase-out as a problem for research, governance and practice and to elicit what is at stake when one tackles phase-out as a long-term process with its corresponding *scales, temporalities, politics and uncertainties*, as well as with a new conception of *risks and how they can be handled*.

4.2 Transitions Studies, Phase-Out, and Different Types of Risk

Having established that abandoning practices or substances that are deemed harmful constitutes a horizon for policy (i.e. phase-out) and for research, we first present a socio-technical perspective on transitions, and then specify some of the implications of such a perspective for the intentional phase-out of socio-technical systems.

4.2.1 Socio-Technical Systems and Transitions as Reconfigurations

Transitions studies (Geels 2002; Köhler et al. 2019; Rip and Kemp 1998) offer relevant approaches to capture the systemic characteristics of socio-technical configurations and problematize the patterns and mechanisms by which these may change. Indeed, this literature suggests examining production-consumption systems as socio-technical configurations, i.e. as bundles of heterogenous entities (scientific knowledge, technology and innovation, production organizations, rules and regulations, users…) that fulfil a particular societal function (e.g. food, shelter, energy, mobility) in a given place and time. Such configurations are understood as being relatively stable over extended periods of time, and hence particularly difficult to change, due to the tight couplings between their constitutive elements and due to their continuous reproduction through practices and institutional logics.

Significantly changing the structure and functioning of such systems hence appears as particularly difficult and rare. Indeed, socio-technical systems tend to absorb injunctions to change through incremental innovation (i.e. marginal and stepwise adjustment of some of its constitutive elements) or substitution (i.e. like-to-like replacement of individual constitutive elements without fundamentally questioning the overall configuration). More rarely, fundamental reconfigurations (i.e. transitions) involve the conjunction of three mechanisms: experimentation in alternative spaces more favourable to radical forms of innovation (socio-technical niches) (Raven et al. 2016; Smith and Raven 2012), the generalization of radical innovations through scaling, diffusion, institutionalization or interaction (Callon 1990; Kanger et al. 2019; Ockwell et al. 2018; Wigboldus et al. 2016; Wilson 2012) and the intentional destabilization of established configurations (Turnheim 2023a). The combination of these mechanisms can generate a variety of transitions patterns and pathways observable in empirical cases (Geels and Schot 2007; Smith et al. 2005).

4.2.2 Governing the Phase-Out

Joly et al. (2023) have further problematized how the phase-out of large sociotechnical systems constitute a particularly tricky challenge, one that calls for the joint consideration of systemic structuration and the temporalities of socio-technical configurations. We here rehearse some of these arguments and specify how they relate to the problem at hand. An essential preliminary step consists in distinguishing (a) the discontinuation of governance from (b) the governance of discontinuation (Stegmaier et al. 2014).

On the one hand, dealing with crises (i.e. governing with the problems and emergencies raised by the exceptional circumstances and urgent temporality of crises and accidents) is essentially a strategic process that resides *outside* of the ordinary modes of governance. Indeed, the main inclination of crisis management, besides avoiding the most harmful and serious damages, is to generate the conditions for a return to the normal, manageable circumstances of governance, i.e. resuming *de facto* governance as swiftly as possible—leaving crisis as a parenthesis to be forgotten.[1] In other words, it is a sort of conservative form of governance that is averse to change with its own institutional dynamic (Ansell et al. 2015) and associated technico-economical forms of lock-in (Arthur 1989).

By contrast, governing discontinuations requires a longer view, one that is specifically targeted at generating the conditions through which entire systems may be phased out, with a particular attention to the depth and far reach of socio-technical entanglements. This requires combining objectives (e.g. long-, mid- and short-term reduction targets) to address harms and risks, interventions explicitly aiming at de-aligning elements of socio-technical configurations so as to enable the detachment of the problematic features (e.g. pesticides dependence, "toxic" food, degraded soils and environments) and support to radically new ways to produce and consume food that alleviate or circumvent risks.

4.3 Illustration: Phasing-Out the Agricultural Use of Pesticides in France

4.3.1 Pesticide Use as Persistent Systemic Problem

From such a perspective, current conventional agri-food systems can be seen as locked-in through commitments to productivism and unsustainable intensification logics (Buttel 2006). The domination exerted by particularly powerful actors and

---

[1] While an important objective of crisis management may be to limit their structural implications in the name of overall stability, such objective cannot always be satisfied—in particular in the case of severe crises and disasters that may require adaptive or transformative change (Ansell and Trondal 2018) or in the case of "creeping" crises characterized by cumulative causation (Boin et al. 2020).

general adhesion to formal and informal rules tend to frame accepted and acceptable practices, which reinforce the "pesticide treadmill" that locks farmers into pesticide dependence (Van den Bosch 1989) and impact the relations between agencies and policy making in the procurement of regulations (Pelaez et al. 2013). More broadly, pesticides reduction or phase-out contributes to the critical agenda of agri-food transitions towards a corporate-environmental food regime (Barbier et al. 2017; Levidow 2015)—understood as the renewal of "strategies for capital accumulation [through the incorporation of] "green" or "quality" products which were previously associated with alternative trajectories from social movements" (Levidow 2015: 86).

Pesticide use in agriculture is a typical form of persistent systemic problem associated to human activity (den Hond et al. 2003; Schuitmaker 2012; Tang et al. 2021). Indeed, since the 1960s, conventional agricultural production has followed a pattern of intensification marked by specialization, economies of scale, and growing dependence upon external inputs (fertilizers, pesticides, mechanization) as well as being place at the crossroad of a sectorial regulation and of more general regulation of food provision in society (Allaire and Boyer 1995). In this regime of intensification, the production, distribution and application of synthetic pesticides for crop protection have become a foundational sub-system upon which food production relies to achieve its current productivity levels. As a consequence, it has become particularly difficult to significantly change pesticide use without fundamentally challenging agri-food systems. In other words, phasing-out pesticide use is a systemic issue calling for systemic approaches to face the challenge of reducing chemical inputs at the level of the agri-food system (Brunelle et al. 2024).

Further, while "pesticides"—a shorthand for "synthetic pesticides"—refer to particular substances resulting from chemical engineering, it is much more than a technical artefact and its tradeable materiality. As such, it calls for a socio-technical perspective to fully appreciate the interlinkages between various sectorial matters of knowledge, innovation, market, health, controversies and regulation:

- the productive apparatus supporting its industrial development,
- the production of scientific and other forms of knowledge on pesticides,
- the innovation around active substances and pesticides product formulations (seeking for novelty of crop protection and blame avoidance or anticipation of molecules phase-out),
- their distribution to various users through specific forms of farming inputs markets,
- the on-site practices of pesticides application and supporting technical knowledge,
- the measurement of accumulation and circulations of active substances through soil, bodies, food and consumers,
- the various forms of knowledge produced and mobilized to support claims of effectiveness and/or toxicity concerning pesticides,
- the regulatory efforts to control or frame the "appropriate" uses of pesticides,
- the opinions and mobilization of various social actors around pesticides issues, as well as the career of controversies and their possible closure.

The governance of (pesticide) phase-out is thus likely to require the combination of explicit discontinuation governance—i.e. the cautious, patient and humble crafting of phase-out pathways and supporting conditions—and more opportunistic seizing of emergent discontinuities such as crises that have the potential to accelerate change if adequately prepared for. But this governance of phase-out calls for critical concerns about the framing relations between science and politics (Boudia and Jas 2015) and the governmentalization of risk through regulatory sciences (Demortain 2011), pointing out that crises not only concern the impact of toxic substances but also the politics through which risks are framed and handled to enact reduction.

4.3.2 Framing Risks, Handling Risks, Enacting Reduction

As far as various actors and publics are concerned, the persistence and systemic attributes of the problems of pesticides use are not undisputable, invariable, or irreversible; they are a matter of framing and referential—themselves constructed, stabilized, fragilized or defended along the definition of public problems. This means that we need to appreciate the emergence of framings and their translations in a longitudinal analysis, and to consider the socio-historical conditions that can explain their (un)felicitous development.

In terms of the development of problem framings, we observe a number of distinct framings at work:

- designating and problematizing pesticides as **dangerous substance**,
- designating and problematizing **dangerous substance** as object of **reduction and removal**,
- designating and problematizing **reduction and removal** as **socio-political risk**.

As our illustration will show, while there is a temporal sequence behind the appearance of these framings, it is essential to resist the urge to appreciate them as linear so as to consider how they interact, become layered and possibly forgotten or replaced.

4.3.3 Governing Phase-Out and Reduction of Pesticides: A Socio-Historical Overview

In this section, we provide a succinct analytical journey through the recent history of pesticides use in agriculture in France, focused on the framing of risks and the attempted handling of reduction challenges in mainstream decision-making settings. We distinguish four periods, which are characterized by relative coherence concerning the dominant handling of pesticides reduction in policy contexts, which tends to privilege certain problem framings and solution pathways over others, as well as opening new marginal spaces for alternatives.

From biocides to the distribution of risks in bodies, environment and food

To understand the prospects of phasing-out pesticide use in the French case, one needs to apprehend it in the long run of a crop protection regime and the way risks have been established therein. The first period is related to the spreading of the US DDT[2] ban in 1972 under the aegis of an environmental movement powerful enough to instigate its withdrawal (Maguire and Hardy 2009). Following the progress of the DDT ban in several countries after the USA,[3] France banned DDT in 1976, although a number of exceptions (namely for curative use) were granted. The early whistle-blowing and environmental claim was clearly established under the auspices of both human and nature protection:

> Can anyone believe it is possible to lay down such a barrage of poisons on the surface of the earth without making it unfit for all life? They should not be called "insecticides" but "biocides". (Carson 1962: 7–8).

While the DDT ban could be seen as a significant landmark in the discontinuation of an insecticide substance, it has to be seen in the perspective of the continuation of the large list of organochlorine and organophosphate pesticides developed immediately in the wake of DDT's proof of concept. In other words, DDT can be seen as a scapegoat substance—allowing its class to remain unharmed (for now…).

Following the concomitant DDT ban and creation of the US EPA, the governmentalization of pesticide uses through norms rested on a sophisticated approach to risk. This approach involved, on the one hand, justification through regulatory science, expertise committee and homologation principles and, on the other hand, a compartmentalization of risks in various problem areas, to be regulated separately: (a) at the level of workers (both in manufacturers and sprayers), (b) at the level of residues in food (the DLL) and (c) at the level of environmental impacts.

During these years, industrial manufacturers of pesticides developed intensive R&D programme to deploy new generations of molecules (such as the systemic pesticides like carbamates) and adapt molecules to homologation constraints. Substitution had thus become an industrial strategy to keep the ambition high with synthetic crop protection products.

In sum, this period saw the designation and problematization of pesticides as dangerous substances to be regulated under dedicated scientific expertise (regulatory toxicology) and administrative procedures (authorization regime), with notable differences between countries to render public contestation and re-stabilization of pesticide use (Levain et al. 2015).

Taming reduction and removal according to risk evaluation

From the 1980s onwards until the Rio Summit in 1992, environmental movements took up the fight against pesticides, resulting in the production of an agenda for the withdrawal of many pesticides. In 1985, the newly-founded Pesticide Action Network

---

[2] DDT is a powerful insecticide used for civil and military purposes, notably to protect crops against insect swarms and to protect human populations against insect-borne diseases like malaria.
[3] Notably, Cuba, Sweden and Norway were early precursors of national DDT bans in 1970.

(PAN) produced a first list of pesticides to be eliminated from the food supply (targeting consumers' health) and ten years after, during its first World Assembly in Vancouver, insisted on the withdrawal of 12 pesticides from human foodstuffs, which included organochlorine/phosphorus compounds, as well as industrial chemicals (HCB and PCB) and industrial byproducts such as dioxins. In relative alignment with the environmental struggles, OECD countries established a specific convention on the withdrawal of persistent organic pollutants, known as the Stockholm Convention, signed in 2001 and implemented by 158 countries by 2004.

Progress with pesticides regulation was largely driven by transnational dynamics impeded by industrial strategies and private governance (Kinniburgh et al. 2023). In fact, it is through European integration that France and a large number of other European countries bound themselves to the implementation of European decisions on assessing the risks posed by plant protection products. A pesticide reduction horizon became part of a political framework with the emergence of a European strategy against pesticides risk, among other toxicity issues (Bozzini 2017).

At the same time, in France and in the rest of Europe, alternative forms of agriculture were emerging, in particular organic farming, along with specific modes of production, processing and supplying that excluded the use of pesticides. The emergence of alternative food systems in an ethical frame of localism (DuPuis and Goodman 2005) signals a form of isolation of chemical pesticide-free agriculture on the fringes of the conventional agricultural production system. The institutionalization of this concrete utopia of a pesticide-free agriculture in farming systems and on specific markets formed a clear socio-technical answer to the withdrawal of pesticide use: radically transforming the cropping systems and the standards of food markets (Goulet et al. 2023).

In sum, his second sequence shows how the end of widely used substances does not mean the end of a socio-technical regime in agricultural production, since the use of pesticides increased during the post-war economic boom (the French "*Trente Glorieuses*") with the extraordinary international growth in agri-food production as part of the green revolution. Concomitantly, the idea of getting rid of pesticides also proved to be feasible and sustainable in the clearly delineated (yet marginal) spaces of organic agriculture (Marsden 2013) and the growing discourses on alternative agri-food movements (Constance et al. 2014). During the 1980s, the agri-food production and distribution system then stabilized around the idea that there could be "bad pesticides" to get rid of and "good pesticides" to keep, as long as their uses complied with the specifications laid down as part of their approval by experts who, at the time, came mainly from the research and industrial R&D departments of the major phyto-pharmaceutical firms.

The socio-political risk and wicked problem of phase-out governance

By the early 2000s in the European context, the brand awareness of organic products and public subsidies for sustainable agricultural practice has accompanied the rise of ecological politics parties and public policies (Lamine and Marsden 2023). National action plans to reduce use started to emerge[4] (Barzman and Dachbrodt-Saaydeh 2011) and initiated the design and implementation of directional public policies to achieve sustainable agriculture in light with nitrogen and pesticide reductions. PAN Europe's critical work at parliamentary level seemed to be bearing fruit. In October 2009, following extensive discussions between the Commission and Parliament, the European Union adopted the "Pesticides Package" (Bozzini 2017), which consists of:

- a regulation that tightens up existing practices for authorizing the marketing of pesticides,
- two separate directives that set out a strategy for reducing the use of pesticides while obliging Member States to draw up national action plans (Directive 2009/128/EC),
- a specific directive on everything to do with regulating the spraying of pesticides, and
- much less anecdotal than it might seem, a regulation obliges Member States to set up a statistical system to monitor domestic consumption and the use of pesticides.

In sum, linking pesticide reduction to sustainable agriculture in a policy referential with mandatory means of implementation had established a new context.

In France, a pesticide reduction horizon becomes part of a political framework with the emergence of this European strategy. Amidst the ongoing environmental battle against pesticides, an increasing number of socio-technical experiments unfolded on the peripheries of conventional agriculture. Moreover, scientific expertise pointed out deterrent effects of pesticides on human and environmental health. Firstly, with the Grenelle agreement of 2007, the reduction strategy spelled out an explicit target for reduction under the first Ecophyto Plan launched in 2008: decrease the use of 50% by the year 2018, "if possible"[5]! Following this momentum, the "Agro-Ecological project for France" of the Ministry of agriculture launched in 2013 reinforced this objective while establishing a clear admonition to conventional agriculture. After the failure of the first Ecophyto plan (Guichard et al. 2017), the second Ecophyto plan (2015) revised the ambition (50% reduction target is maintained but pushed back to 2025, with an intermediate objective of 25% reduction set for 2020) and stressed that the reduction of pesticides must take place in conventional farming, with the emergence of a form of gradual conventionalization of organic farming and the experimentation of agro-ecological farming practices. Hence, the risk of pesticide use turned out to be political: would it be possible to generalize pesticide reduction with incentives and objectives addressed to the whole agricultural sector? Could such a policy combining pesticide-free agriculture and pesticide reduction be consistent? From today, the answer is no, as the existence of the third Ecophyto Plan

---

[4] National Pesticides reduction plans were launched in Denmark in 1986, in Sweden in 1986 and in The Netherlands in 1991.

[5] How the qualification "if possible" was added later in the process (see Dedieu 2022) and denotes a depression of initial objectives by tying them explicitly to economic bottom lines and technological alternatives (i.e. appropriate substitutes).

issued in 2018 (or so-called Plan Ecophyto 2 +) deals with the concerning end of the glyphosate (a new scapegoat?), but largely prolonged the same objectives and means with the already declared targets of research and innovation efforts, stewardship for change and prevention towards exposure.

The succession of these plans affords for the existence of a socio-political lock-in: three successive Ecophyto reduction plans have failed to reign in pesticide use, which has even increased after the announcement of the first plan. The reason of that failure has been documented, studied and analysed (Dedieu 2022) and it has become a public proof of deep prevailing lock-ins, both technical and economical, linking chemical crop protection efficiency to farm performance (Potier 2023, 2014). The recent farmers' movement in the Netherlands, Germany and France has ostensibly shouted with tractor convoys that the European system of environmental and health norms, among which pesticide reduction, was putting farmers at risk of closing-down their activity—if not committing suicide, while the farmers' poisoning by pesticide use are known to be under-reported (Dedieu et al. 2015). The idea of phasing-out pesticide use, or even the objective of pesticide reduction seems to have faded away in a never-ending horizon.

The question of keeping the ambition high with the Green Deal is similarly on the agenda at the beginning of 2024. In the face of the risk of economic collapse, the French "thermometer" of pesticide reduction has changed from the NODU[6] to European index HRI-1.[7] HRI-1 clearly values the dangerousness of pesticide and not substances units release independently. Meanwhile, the French government has abandoned its ambition of reducing pesticide use by 50% by 2030. In this last sequence, while agro-ecology for France initially appeared as a strong critical alternative, it is now associated with the definition of an ever-postponed reduction horizon. The French case is thus emblematic of a dual political effort, both at the level of those involved in ecological criticism and at the level of public policy "to reduce, if possible". The parenthesis of the agro-ecology project for France seems to have been closed, all the more so since a new parliamentary report highlights the failure to reduce the use of pesticides, and there is now talk of 2050 as the deadline for achieving such a reduction.

---

[6] The number of dose units of a substance per ha (NODU) measures the normalized quantity of an active substance obtained by dividing the quantity sold of each active substance by the "maximum dose" specific to each active substance that can be applied during an "average" treatment on a crop in a given year.

[7] This harmonized risk indicator for pesticides is calculated on the basis of sales of plant protection products, taking into account the toxicity of the molecule, it measures the sum of the quantities of active substances sold in a year, weighted by a coefficient that can take four different values depending on the toxicological classification of the active substance.

4.4 Conclusion

A socio-technical and longitudinal transitions perspective on the (non) phase-out of pesticide use in agriculture allows us to shed light on some immediate features of such problem. The involvement of multiple actors and actor positions concerning phase-out is a matter of various types of mobilizations showing oscillations between the ontological claim of preserving various forms of life and the ontological claim of maintaining business-as-usual. The scientific arena is not positioned outside (above or under) the manufacture of various risks (worker, food, environment); it is the same for policy-making arenas (at various levels) of dealing with those risks as regulatory targets and with the procedures of putting the regulation at work through substantial sectorial changes. The socio-historical overview of the French case clearly shows an accumulation of framings at work, as pesticide phase-out or reduction appears to be loaded gradually with problematizations which are embedded in mobilizations about: (a) definitional struggles to designating and problematizing pesticides as dangerous substance; (b) procedural struggles to designating and problematizing dangerous substance as object of reduction and removal; and (c) directional governance to designating and problematizing reduction and removal as a socio-political risk.

But this process is far from being linear and cumulative since numerous questions of inertia, heritage and temporality are systematically at play to define and enact the condition of possibilities of a phase-out and reduction. Agri-food systems can be approached as systemic configurations held together by the tight couplings between heterogenous problematizations of phase-out and reduction.

New problems and concerns (e.g. raised awareness and problematization of toxicity and health impacts pesticides) have certainly triggered the partial delegitimation of conventional agri-food practices and opened up new avenues for innovation; but our analysis also shows how intentional governance has a tendency to be folded back into the seemingly inescapable and overbearing constraints of "feeding the world"—in a particular formulation that currently appears impermeable to critique.

Removing pesticides from agricultural use through incremental or substitution logics (e.g. replacement by biopesticides) appears likely to be ineffective or marginal. More astonishing perhaps are the backlashes and the constant survival of the substitution of molecules framework that tend to close down the framing in terms of phase-out and risk reduction. Removing individual phytosanitary substance(s) (classes) (e.g. DDT, glyphosate or nicotinoids) is unlikely to fundamentally alter agri-food configurations and their related dependence upon external inputs.

In sum, explorations of the possibility of phasing-out pesticides have unveiled what we have called a socio-political risk of not fulling the promise of opening a new era for agri-food systems. This confirms—in our view—that pesticide phase-out needs to be approached at a systemic level, in conjunction with support to radically alternative pathways for agri-food system development (e.g. the large-scale deployment of agro-ecological practices). Phasing-out pesticides, like dealing with other kinds of persistent systemic issues, appears as if imbued with a still currently unbearable heaviness. Confronted with what appears as the impossible task of phasing-out undesirable (risky) systems while ensuring that such (risky) ends are not undesired, it becomes clear that what is needed is an appropriate framing of the new kinds of risks associated with phase-out. Having established what such a new framing entails, it becomes clear that this requires anticipating and frontally addressing the sociopolitical risks of phase-out, rather than leaving them as unattended afterthoughts. With this, phase-out may be bearable.

References

G. Allaire, R. Boyer, *La grande transformation de l'agriculture: lecture conventionnalistes et régulationistes* (Inra-Quae, Paris, 1995)

C. Ansell, A. Boin, M. Farjoun, Dynamic conservatism: how institutions change to remain the same, in *Institutions and ideals: Philip Selznick's legacy for organizational studies* (Emerald Group Publishing Limited, 2015), pp. 89–119

C. Ansell, J. Trondal, Governing turbulence: an organizational-institutional agenda. Perspect Public Manag. Gov. **1**, 43–57 (2018)

W.B. Arthur, Competing technologies, increasing returns, and lock-in by historical events. Economics J. **99**(394), 116–131 (1989)

M. Barbier, B. Elzen, B. van Mierlo, A.M. Augustyn, 2017. Conclusion: a curiosity cabinet of agroecological transition studies, in: *AgroEcological Transitions: Changes and Breakthroughs in the Making* ed. by B. Elzen, A.M. Augustyn, M. Barbier, B. Van Mierlo (Wageningen University, pp. 283–295)

M. Barzman, S. Dachbrodt-Saaydeh, Comparative analysis of pesticide action plans in five European countries. Pest Manag. Sci. (2011). https://doi.org/10.1002/ps.2283

A. Boin, M. Ekengren, M. Rhinard, Hiding in plain sight: conceptualizing the creeping crisis. Risk Hazards Crisis Public Policy **11**, 116–138 (2020)

S. Boudia, N. Jas, Introduction: science and politics in a toxic world, in: Toxicants, Health and Regulation since 1945 (Routledge, 2015), pp. 1–23

E. Bozzini, *Pesticide policy and politics in the European Union: regulatory assessment, implementation and enforcement* (Springer, 2017)

T. Brunelle, R. Chakir, A. Carpentier, B. Dorin, D. Goll, N. Guilpart, F.H. Tang, Reducing chemical inputs in agriculture requires a system change. Commun. Earth Environ. **5**(1), 369 (2024)

F.H. Buttel, Sustaining the unsustainable: agro-food systems and environment in the modern world, in *Handbook of Rural Studies* (2006), pp. 213–229

M. Callon, Techno-economic networks and irreversibility. Sociol. Rev. **38**, 132–161 (1990). https://doi.org/10.1111/j.1467-954x.1990.tb03351.x

R.L. Carson, *Silent spring* (Houghton Mifflin Company, Boston, 1962)

D.H. Constance, W.H. Friedland, M.C. Renard, M.G. Rivera-Ferre, The discourse on alternative agrifood movements, in *Alternative Agrifood Movements: Patterns of Convergence and Divergence* (Emerald Group Publishing Limited, 2014), pp. 3–46

F. Dedieu, J-N. Jouzel, G. Prete, Governing by ignoring: the production and the function of the under-reporting of farm-workers' pesticide poisoning in French and Californian regulations, in *Routledge International Handbook of Ignorance Studies* (Routledge, 2015), pp. 297–307

F. Dedieu, Pesticides: le confort de l'ignorance, Seuil (2022)

D. Demortain, Scientists and the regulation of risk: standardising control (Edward Elgar Publishing, 2011)

E.M. DuPuis, D. Goodman, Should we go "home" to eat?: toward a reflexive politics of localism. J. Rural. Stud. **21**, 359–371 (2005)

F.W. Geels, Technological transitions as evolutionary reconfiguration processes: a multi-level perspective and a case-study. Res. Policy **31**, 1257–1274 (2002). https://doi.org/10.1016/S0048-7333(02)00062-8

F.W. Geels, J. Schot, Typology of sociotechnical transition pathways. Res. Policy **36**, 399–417 (2007)

F. Goulet, A. Aulagnier, E. Fouilleux, Moving beyond pesticides: exploring alternatives for a changing food system. Environ. Sci. Policy **147**, 177–187 (2023)

L. Guichard, F. Dedieu, M.-H. Jeuffroy, J.-M. Meynard, R. Reau, I. Savini, Le plan Ecophyto de réduction d'usage des pesticides en France : décryptage d'un échec et raisons d'espérer. Cahiers Agricultures **26**, 14002 (2017). https://doi.org/10.1051/cagri/2017004

F. den Hond, P. Groenewegen, N.M. van Straalen, Questions around the persistence of the pesticide problem, in *Pesticides: Problems, Improvements, Alternatives* (Wiley, 2003), pp. 1–15. https://doi.org/10.1002/9780470995457.ch1

P-B. Joly, M. Barbier, B. Turnheim, Governing the discontinuation of large socio-technical systems, in F. Goulet, D. Vinck, ed. by *New Horizons for Innovation Studies: Doing Without, Doing With Less*. Edward Elgar (2023), pp. 29–46. https://doi.org/10.4337/9781803925554

L. Kanger, F.W. Geels, B. Sovacool, J. Schot, Technological diffusion as a process of societal embedding: lessons from historical automobile transitions for future electric mobility. Transp. Res. D. Transp. Environ. **71**, 47–66 (2019). https://doi.org/10.1016/j.trd.2018.11.012

F. Kinniburgh, H. Selin, N.E. Selin, M. Schreurs, When private governance impedes multilateralism: the case of international pesticide governance. Regul. Gov. **17**, 425–448 (2023)

J. Köhler, F.W. Geels, F. Kern, J. Markard, A. Wieczorek, F. Alkemade, F. Avelino, A. Bergek, F. Boons, L. Fünfschilling, D. Hess, G. Holtz, S. Hyysalo, K. Jenkins, P. Kivimaa, M. Martiskainen, A. McMeekin, M.S. Mühlemeier, B. Nykvist, B. Pel, R. Raven, H. Rohracher, B. Sandén, J. Schot, B. Sovacool, B. Turnheim, D. Welch, P. Wells, An agenda for sustainability transitions research: State of the art and future directions. Environ. Innov. Soc. Transit. **31**, 1–32 (2019). https://doi.org/10.1016/j.eist.2019.01.004

Z. Koretsky, P. Stegmaier, B. Turnheim, H. van Lente, (ed.), 2023. *Technologies in Decline: Socio-Technical Approaches to Discontinuation and Destabilisation*. (Routledge)

C. Lamine, T. Marsden, Unfolding sustainability transitions in food systems: insights from UK and French trajectories. Proc. Natl. Acad. Sci. **120**. (2023) https://doi.org/10.1073/pnas.2206231120

L. Levain, PB. Joly, M. Barbier, V. Cardon, F. Dedieu, et al., Continuous discontinuation—the DDT Ban revisited. International Sustainability Transitions Conference "Sustainability transitions and wider transformative change, historical roots and future pathways", (University of Sussex. Brighton, GBR., Brighton, United Kingdom 2015)

L. Levidow, European transitions towards a corporate-environmental food regime: agroecological incorporation or contestation? J. Rural. Stud. **40**, 76–89 (2015). https://doi.org/10.1016/j.jrurstud.2015.06.001

S. Maguire, C. Hardy, Discourse and deinstitutionalization: the decline of DDT. Acad. Manag. J. **52**(1), 148–178 (2009). https://doi.org/10.5465/amj.2009.36461993

B. Mansfield, M. Werner, C. Berndt, A. Shattuck, R. Galt, B. Williams, L. Argüelles, F.R. Barri, M. Ishii, J. Kunin, P. Lapegna, A. Romero, A. Caicedo, Abhigya, M.S. Castro-Vargas, E. Marquez, D. Ojeda, F. Ramirez, A. Tittor, A new critical social science research agenda on pesticides. Agric Human Values **41**, 395–412 (2024). https://doi.org/10.1007/s10460-023-10492-w

T. Marsden, From post-productionism to reflexive governance: contested transitions in securing more sustainable food futures. J. Rural. Stud. **29**, 123–134 (2013). https://doi.org/10.1016/j.jrurstud.2011.10.001

D. Ockwell, R. Byrne, U.E. Hansen, J. Haselip, I. Nygaard, The uptake and diffusion of solar power in Africa: socio-cultural and political insights on a rapidly emerging socio-technical transition. Energy Res. Soc. Sci. **44**, 122–129 (2018). https://doi.org/10.1016/j.erss.2018.04.033

V. Pelaez, L.R. da Silva, E.B. Araújo, Regulation of pesticides: a comparative analysis. Sci. Public Policy **40**, 644–656 (2013). https://doi.org/10.1093/scipol/sct020

D. Potier, Pesticides et agroécologie, les champs du possible (2014)

D. Potier, Rapport pour la création d'une Commission d'enquête sur les causes de l'incapacité de la France à atteindre les objectifs des plans successifs de maîtrise des impacts des produits phytosanitaires sur la santé humaine et environnementale et notamment sur les conditions de l'exercice des missions des autorités publiques en charge de la sécurité sanitaire (2023)

R. Raven, F. Kern, B. Verhees, A. Smith, Niche construction and empowerment through socio-political work. a meta-analysis of six low-carbon technology cases. Environ. Innov. Soc. Transit **18**, 164–180 (2016). https://doi.org/10.1016/j.eist.2015.02.002

A. Rip, R. Kemp, Technological Change, in *Human Choice and Climate Change*, ed. by S. Rayner, L. Malone. Resources and Technology, vol. 2 (Batelle Press, Washington, D.C., 1998), pp.327–399

T.J. Schuitmaker, Identifying and unravelling persistent problems. Technol. Forecast Soc. Change **79**, 1021–1031 (2012). https://doi.org/10.1016/j.techfore.2011.11.008

A. Smith, R. Raven, What is protective space? Reconsidering niches in transitions to sustainability. Res. Policy **41**, 1025–1036 (2012). https://doi.org/10.1016/j.respol.2011.12.012

A. Smith, A. Stirling, F. Berkhout, The governance of sustainable socio-technical transitions. Res. Policy **34**, 1491–1510 (2005). https://doi.org/10.1016/j.respol.2005.07.005

P. Stegmaier, S. Kuhlmann, V.R. Visser, et al., The discontinuation of socio-technical systems as a governance problem, in *The Governance of Systems Change: Explaining Change*, ed. by J. Edler, S. Borrás, pp. 111–131 (2014)

F.H.M. Tang, M. Lenzen, A. McBratney, F. Maggi, Risk of pesticide pollution at the global scale. Nat. Geosci. **14**, 206–210 (2021). https://doi.org/10.1038/s41561-021-00712-5

B. Turnheim, Destabilisation, decline and phase-out in transitions research, in *Technologies in Decline: Socio-Technical Approaches to Discontinuation and Destabilisation*, ed. by Z. Koretsky, P. Stegmaier, B. Turnheim, H. Van Lente (Routledge, (2023a)

B. Turnheim, The historical dismantling of tramways as a case of destabilisation and phase-out of established system, in *Proceedings of the National Academy of Sciences*, vol. 120 (2023b). https://doi.org/10.1073/pnas.2206227120

R. Van den Bosch, *The pesticide conspiracy* (University of California Press, 1989)

S. Wigboldus, L. Klerkx, C. Leeuwis, M. Schut, S. Muilerman, H. Jochemsen, Systemic perspectives on scaling agricultural innovations a review. Agron. Sustain. Dev. **36** (2016). https://doi.org/10.1007/s13593-016-0380-z

C. Wilson, Up-scaling, formative phases, and learning in the historical diffusion of energy technologies. Energy Policy **50**, 81–94 (2012). https://doi.org/10.1016/j.enpol.2012.04.077

Open Access This chapter is licensed under the terms of the Creative Commons Attribution 4.0 International License (http://creativecommons.org/licenses/by/4.0/), which permits use, sharing, adaptation, distribution and reproduction in any medium or format, as long as you give appropriate credit to the original author(s) and the source, provide a link to the Creative Commons license and indicate if changes were made.

The images or other third party material in this chapter are included in the chapter's Creative Commons license, unless indicated otherwise in a credit line to the material. If material is not included in the chapter's Creative Commons license and your intended use is not permitted by statutory regulation or exceeds the permitted use, you will need to obtain permission directly from the copyright holder.

Chapter 5
Glory, Mourning, Memory. Archiving Knowledge, Dismantling Nuclear Power

Christine Bergé

Abstract Deconstructing, dismantling, closes what was founded. In the cycle of human institutions, the end of life shall enter into a controlled process. It is not a question of destroying or letting it fall into ruin, but finding the path to a new legacy, after a period of mourning. For several centuries, numerous "objects" (architectures, vessels, hospitals, libraries, etc.) have undergone closing-down or remodelling, and no specific term has designated this action. Why have these processes, through the term *dismantling*, now acquired a new dimension? Between obsolescence, dismantling and new modalities of memory, let's follow in this chapter the destiny of three technological objects.

Keywords Dismantling · Institution · Nuclear plant · Museum · Archive knowledge

5.1 Introduction

Our horizon has changed. We are caught between obsolescence (of technologies, of our world) and the desire to adapt to change. We sense that we are entering a grey zone where our survival is not assured if we do not invent new ways of living. Our legacies are both problematic and controversial.

I will compare two different fields, which question the ending of institutions considered as technical beings (cultural devices, industrial tools) and quasi-organic ones (their birth and death involve more than questions of functionality). How to rethink knowledge institutions? These places where specific knowledge is developed and transmitted (such as hospitals, museums, libraries, universities, industries), how are they transforming to adapt to social changes? What is the lifespan of an institution considered as a tool for transmitting values and knowledge?

I will take the example of the deconstruction of the natural history museum in Lyon (France) and its refoundation into two knowledge institutions which revalue

---

C. Bergé (✉)
Paris, France
e-mail: berge.christine@free.fr

© The Author(s) 2025
M. Bourrier (ed.), *Decommissioning Aging Installations and Declining Technologies*, SpringerBriefs in Safety Management, https://doi.org/10.1007/978-3-031-88369-9_5

and renew the memory of the knowledge they preserve. The conservation of archived objects is an important issue here. Regarding the dismantling of industrial tools, I will take the example of Superphénix, a nuclear power plant, and the archiving of its materials: in this case, technical invention faces safety concerns, while long-term waste is archived, and other waste is recycled.

These areas could be introduced by the analysis of a technical object: Kanak headbreakers, weapons of war which, uprooted from their ecological niche by travellers, lost their meaning and their memory upon arriving in our museums. How can we document and reconstruct these objects? Comparisons show us that future paths can be different, even if the transversal issues persist: the glory of growth, the transition from mourning to refoundation, and the long timescales of human memory.

5.2 Museography: Analysis and Reconstruction of a Cultural Object

For a century and a half, anthropology and museography have been safeguarding and studying cultural heritage. How to understand the role and meaning of objects that arrive in museums or private collections? Ethnographic work has been capable of approaching part of their native living conditions. They had their meaning within rituals, knowledge, and daily practices which transmitted their heritage value, sometimes for several millennia. The objects had material bodies and immaterial values. Unfortunately, to share visions of the world from several cultures, the museum strips objects from their original *niche* to insert them into educational staging practices intended to present them to the public. These objects lose their memory behind the glass where their naked form is deprived of the gestures that nourished them. Is the museum a kind of cemetery where mourning for once-living objects takes place? Not necessarily. The exchanges between exhibition sites and ethnographic observation have been able to enrich the understanding of these objects. Techniques for analysing and reconstructing forms and meaning allow us to perceive their complexity. The museum becomes a place of archiving "in the meantime", where we can return several times to study the beings extracted from their original *niche*, and give them back a part of their memory, as we will see with the example of Kanak headbreakers. Emmanuel Kasarhérou, president of the Quai Branly Museum in Paris, worked on the inventory of dispersed Kanak heritage, for an exhibition in 2013 in the museum. He collected considerable photographic documentation relating to Kanak objects in several museums in France and Europe. The anthropologist Christian Coiffier, lecturer at the National Museum of Natural History in Paris, testifies to the fruitfulness of such an enterprise in terms of cultural exchanges.[1] These actions allow us to question the notion of "heritage": what is inherited from ancestors takes on its meaning in a territory, through indigenous human practices approached by Coiffier.

---

[1] C. Coiffier is also responsible for the Research and Teaching Department of the Musée du Quai Branly.

How then can such heritage take on meaning in another culture, when presented in a museum? In the hands of chiefs, Kanak headbreakers were glorious weapons of war. "In the nineteenth century" writes Coiffier,

> Kanak headbreakers frequently presented various plant wrappings on their handles with, sometimes, the addition of bouquets of plants tied to them. The insufficient maintenance of these objects in their places of Conservation has too often, over the past century and a half, led to the destruction of these meaningful plant elements. The latter do not seem to have been of much interest to collectors and museum curators who focused more on the shape of the objects. The various plant elements have too often been perceived as simple decorations so that there is very little documentation about them. It is too often the material function of the object that has been retained rather than its immaterial function.
>
> From Coiffier 2013, *Fougères et autres éléments végétaux associés aux casse-têtes kanak ou l'art de communiquer sans parole*, Journal de la Société des Océanistes.[2] Reprinted with permission.

In European museums, scrupulous curators have saved the headbreakers with their plant parts. Coiffier observed them and noted the presence of a small recurring number of plant species, especially ferns, of which one or two specific types had been selected by the Kanak. Coiffier tends to demonstrate the existence of a language of signs coded through figures woven into these fibres, to ensure communication between several populations who did not always speak the same language. Even if it is perhaps "too late to understand their true meanings", the anthropologist shows that the way in which such a plant was tied to the pole of the headbreakers had a mnemonic function. The meaning inscribed in the intertwining, "every initiated man had to understand it, even beyond his own linguistic area".

The first Kanak headbreakers brought back to Europe during Captain Cook's second expedition to the Pacific had already been stripped of their plants. According to Coiffier, the ethnographic collections were made up of Kanak weapons for close combat, such as the headbreakers which were used to break the vertebrae of the neck. Many of these weapons were confiscated or banned from use in a large part of New Caledonia at the end of the nineteenth century and replaced... by rifles! Headbreakers became exotic objects hunted by collectors. Coiffier describes the shapes: straight headbreakers with round, flattened, star or half-ball heads, with grooves, headbreakers with acorn or mushroom-shaped heads, turtle head or bird head headbreakers, sickle headbreakers; all made of very hard wood ("iron tree"), wrapped in bark fabrics (tapa) or palm fibres tied using different techniques. The interior parts of the bark (bast) used had a particular meaning: thus, the white tapa was used to make penis sheaths and turbans, as described again by Coiffier:

> This sheathing could be made over the entire handle of the headbreakers or only at its end. It was fixed using ligatures made from cords of red-dyed flying fox hair or coconut fibers.

---

[2] Coiffier 2013, *Fougères et autres éléments végétaux associés aux casse-têtes kanak ou l'art de communiquer sans parole*, in "La part d'immatériel dans la culture matérielle", Journal de la Société des Océanistes, p. 133–147. https://doi.org/10.4000/jso.6989.

From Coiffier 2013, *Fougères et autres éléments végétaux associés aux casse-têtes kanak ou l'art de communiquer sans parole*, Journal de la Société des Océanistes.[3] Reprinted with permission.

By exploring ethnographic documents, Coiffier restores the relationship between technique and symbolism. He shows how materials and gestures of making the headbreakers integrate within themselves the cultural environment of the objects and the Kanak institutions (religious, political, matrimonial) which formed the world where these objects took on meaning. He analyses the system of "ligatures, either crossed so as to form diamonds, or wound into a spiral" which weaved these plants tied to the headbreakers. The insertion of specific objects (fine basketwork rings, amulets, etc.) carried specific meanings in themselves. Many headbreakers had only a ritual function. Deciphering these objects thus reveals their links to Kanak institutions. The actors read these shapes and materials which informed them about the clan affiliation and rank of the person wearing the object. They contained mnemonic elements concerning ecological, meteorological, medicinal and food knowledge, as well as matrimonial rules, ritual poems.

It was a non-verbal communication process. The ferns tied to the headbreakers were therefore part of the carefully wrapped hidden elements, in certain cases, or exposed for public view in other cases:

> They only represented one term of the message, the type of material used to fix them, its texture, its color and the technique of the knots made must have been other relevant elements for a good understanding.
>
> From Coiffier 2013, *Fougères et autres éléments végétaux associés aux casse-têtes kanak ou l'art de communiquer sans parole*, Journal de la Société des Océanistes.[4] Reprinted with permission.

Deconstructing the object to analyse it opens up clues to its original meaning. Christian Coiffier promotes inter-cultural exchange and attempts to translate/transform a foreign heritage into universal heritage. In this "reconstruction", the object finds a memory which makes the link with indigenous institutions, their inheritance.

---

[3] Coiffier 2013, *Fougères et autres éléments végétaux associés aux casse-têtes kanak ou l'art de communiquer sans parole*, in "La part d'immatériel dans la culture matérielle", Journal de la Société des Océanistes, p. 133–147. https://doi.org/10.4000/jso.6989.

[4] Coiffier 2013, *Fougères et autres éléments végétaux associés aux casse-têtes kanak ou l'art de communiquer sans parole*, in "La part d'immatériel dans la culture matérielle", Journal de la Société des Océanistes, p. 133–147. https://doi.org/10.4000/jso.6989.

5.3 An Institution Between Deconstruction and Refoundation: The Natural History Museum of Lyon

In 1878, work began on the first Guimet museum in Lyon. The industrialist and collector Émile Guimet wanted to build an institution open to the public, to house a religious museum and "put all the gods under the same roof". Refined rooms, sophisticated lighting appealed to the most cultured. Twenty years later, the number of visitors declined, and the business collapsed. The Société Frigorifique de Lyon ended up acquiring the premises in 1901. The Lyon Museum moved away, and the premises were transformed into an industrial complex dedicated to the manufacture of ice, with the installation of an ice rink: it was the *Palais des Glaces*, soon to be enriched with a restaurant, a theatre and a winter garden. After its hour of glory, the audience dwindled. In 1909, the Société Frigorifique filed for bankruptcy. The city of Lyon bought the premises, Mayor Édouard Herriot had the ice rink dismantled to create in its place the "great room" (see Fig. 5.1) of the Natural History Museum of Lyon, which was opened in 1914.

The old Guimet museum had now lived three lives. In the imposing architecture then developed an archive of animal skeletons and naturalized creatures, a Noah's ark which became the place for sending and donating private collections, for exchanges between researchers, with its kilometres of shelves of scholarly works, its display cases, cabinets, book drawers and jars enclosing thousands of fascinating specimens. Snakes, spiders, winged or non-winged insects, butterflies, birds, bats, bears, primates, human skeletons, prisoners' skulls, Egyptian mummy heads, as well as

Fig. 5.1 The "Grande Salle" of the Lyon Natural History Museum. Copyright 2002 by Jacqueline Salmon. All rights reserved

numerous human artefacts, provided the material for exhibitions which were open to the public during the twentieth century. The collections were the pride of the Museum, celebrating the sun of knowledge which never sets on the globe. In addition to the classification and archiving of these "objects", problems related to conservation had to be resolved: technologies, with their own history.

In 2000, the General Council of the Rhône gave *carte blanche* to the photographer Jacqueline Salmon to create the "memory" of the know-how of the MHN of Lyon. As the conditions of conservation and research have evolved, it appeared necessary to redefine the cultural and scientific project of the institution. The museum had to withdraw from public life to meditate on its metamorphosis into a Center for the Study and Conservation of Collections, and a Museum of "World Cultures". This "memory" of the MHN was to close an era. During the time when the places were deserted, before the move, I worked on this project alongside my photographer friend who suggested that I write the text of our book *Natural Archives*, based on interviews with the researchers who study scientific collections. Later, contemporary architecture would be revealed, a multifaceted crystal at the confluence of the Rhône and the Saône: the Confluences Museum. Part of an international exchange movement, it also houses old collections in state-of-the-art technical and scientific conditions.

I opened the work under the auspices of an "Ethnography of heritage", the term heritage already showing all its complexity. Everything that the Earth sciences preserved at the MHN still bore the marks of their extraction, making legible the "archive" effect of the earth itself (fossil clays, quarries) as a place where the strata may be deciphered in future. Thus, the journey of an ammonite, from an ancient Spanish sea, in the suitcases of a researcher who, after having studied it, suddenly asked the question: why had he not found its base, its imprint, in its original environment? He returned to the location and noticed that a collapse on the ground had separated and shifted the ammonite several metres from its mineral support, which he was able to extract and bring back.

Sciences of Man, Life and Earth, at the end of the nineteenth century, enjoyed an idea of knowledge tools, to approach a profusion of life to be deciphered. The so-called reference collections were part of vast networks of scientific exchange. This is still the case today. But living things seem more fragile to us, the techniques for archiving elements to be preserved are riddled with questions, new links are being forged between the public gaze and the threatened environment. While browsing the material archives of the MHN currently being deconstructed, we came across the remains of "extinct" species. The issue of biodiversity loss (local, global) was already acute in the minds of researchers. One of the entomologists was passionate about the smallest ladybug in France, *Cylindera arenaria*, which lived among other places on the banks of the Rhône and its tributaries, but the embankment works, canals, dams, dredging caused its loss.

The collections were therefore going to move. For example, I observed the techniques of stitching, conservation, archiving, and the fabulous books composed of insects, legs and wings spread, aligned like writing signs, which presented themes and variations of shapes around a specimen. In the "reserves", huge drawers contained birds "in skin": their bodies emptied of their perishable substance, clean and dry, wrapped in white tissue paper, they awaited the moment of their reconstitution. Sometimes, at the bend of a corridor, a "naturalized" (artificialized) animal waited in front of the door of a laboratory where it would soon undergo a full examination of its state of conservation, before its packaging and transport.

Should we mourn the concept of nature? We need to welcome new concepts like that of "co-evolution", which puts us back in our place as living beings among others. It supports the change of view towards other cultures, in favour of new ways of thinking about transculturalism by involving more of the presence of indigenous people. Thus, deconstructed, the old museum lost its designation as "natural history". The institution was reborn through the conceptual renewal of museography. This makes us perceive the fragility of our ecosystems. Inheritance therefore requires creating new institutional approaches.

5.4 What Does It Mean to "Dismantle" Nuclear Power?

The verb "dismantle" comes from the vocabulary of war, and means destroying fortifications, tearing down the walls of a fort or the ramparts of a city. Its application to the industrial environment involves methodically dismantling it, right down to the foundations of architecture. In Creys-Malville, the dismantling of Superphénix has been underway since 1998. It was once again Jacqueline Salmon, the photographer, who introduced me, in 2008, to this site which she was documenting: the site opened an exhibition on "ten years of deconstruction". I wrote the text that would accompany the photos, which were published in two volumes (Bergé 2010, Bergé and Salmon 2011).

November 2008: upon arriving at the site, I noticed the clocks in the buildings, hidden by white cardboard. Suspended time is, symbolically, that of deconstruction: operations always risk being more complex than expected. The time T for reactor core shutdown is contained in four letters: MHSD, permanent decommissioning.[5] This "cardiac" arrest implies the immediate opening of specific locations, as can be seen in the photo in Fig. 5.2. Two temporalities will coexist for some time: that of inert materials to be deconstructed, and that of liquid sodium, Superphénix being an fast neutron reactor (FNR) which used sodium as its heat transfer fluid. This radioactive sodium, treated drop by drop to be archived in concrete cubes measuring one metre on a side, was to remain liquid until the end of the process in 2014. The reactor organism can be thought of as a body whose vital functions are cut according to different rhythms.

It was in Marcoule in 1973 that Phénix, a sodium-cooled fast neutron reactor, was started. The company European Fast Neutron Nuclear Power Plant, a limited company (NERSA), created the same year, brought together the economic investments of six European countries. It was responsible for building the Superphénix

---

[5] MHSD is the French acronym for Mise Hors Service Définitive. Also: MAD (Mise à l'Arrêt Définitive, a permanent shutdown). It authorizes the dismantling of the reactor core.

Fig. 5.2 Top view of the reactor vessel. Copyright Jacqueline Salmon. All rights reserved

reactor, based on the same model, but on an industrial scale, which was to be the prototype of a new sector. The "elders" testify to their enthusiasm during this golden age of French nuclear power. For them, Superphénix was the most powerful breeder reactor. They felt a bit like the alchemists of ancient times. They spoke with nostalgia of the time when the power plant was operating at full capacity.

The moment of glory lasted a few years, and then recurring technical problems led to the shutdown of the plant in 1997, by ministerial decision. The elders experienced the stopping of this "young man's heart" as a technical murder. They organized a symbolic burial for him and began a long mourning. Then they began to dismantle the vital functions of their work tool.

The dismantling process now entered a slow history, in a transnational technical culture that was fed by the exchange of operating experience from other facilities (called "retour d'expérience" or REX in French). Indeed, part of the Dounreay RNR in Scotland, also in the process of deconstruction, participated in these knowledge exchanges. The work in progress had few reverse movements: in 2003, on the Creys-Malville site, some refused to "break" the engine room. There was still one year left to finish "unloading" the reactor core, which was shut down on December 30, 1998. This operation required specific techniques, waiting for authorizations and validations from commissions, under the direction of the Centre Lyonnais d'Ingénierie (CLI), former EDF headquarters, then Deconstruction and Environmental Engineering Center (EDF CIDEN) created in 2001. The "elders" thought they would have to start all over again.[6]

---

[6] The idea of a partial dismantling of nuclear power plants, followed by a 40-year wait, was abandoned.

An automated handling line was created. In December 1999, the extraction of the irradiated assemblies from the reactor core began, which were to be transported on site to the workshop pool for storage.[7] This workshop, built in 1989, was used to store the irradiated core as well as the assemblies of the second core, which had never been used. During the transfer, each assembly (with its number, positioned at a precise place in the map of the reactor core) was removed from the "bed base", and then accompanied by a card which memorized its journey throughout the process. As an engineer explained, the assemblies passed blindly, with a camera and detectors, thanks to automatons controlled from the control room. It must be emphasized that such dismantling is only possible in the case of a power plant whose core has never melted.

Once removed from the tank, each assembly was wrung out then lowered into a washing well to be freed of its residual sodium. This highly radioactive sodium then had to undergo the slow process of carbonation, a process developed within the CEA and used in the Phénix power plant: Superphénix benefited here from experience feedback from the earlier work.[8] Each assembly was washed with water in the wash well, after which the water had to be treated. At the same time, the assemblies were reassembled in the handling cells. The assembly was finally taken into a specific protective enclosure, in order to reach the final location reserved for it in the pool for combustible storage.

We see, with this example, that special workshops had to be installed to carry out the dismantling of the core. In 2007, EDF did not consider these elements as waste, but as fuel reserves that could later be reused. Once installed in its "rack" in the spent fuel pool, the radioactive radiation and the residual heat of each assembly could be attenuated; the water had to be monitored continuously. Disconnect, dismantle, decontaminate, cut: all the operating elements of the plant must undergo these operations (still in progress today) which require the invention of *ad hoc* technologies, some of which are illustrated in Fig. 5.3. Disconnection concerns the separation of the vital function, dismantling refers to the following operations, carried out by specialized actors. The elements are then archived according to their degree of radioactivity: for example, FAMA waste (low to medium activity) must receive approval for its transport to Soulaines (Aube, France), at the time the largest storage centre of radioactive waste on the surface in the world, in operation since 1992.[9]

The elements resulting from nuclear deconstruction follow two paths: recycling and archiving. The issues related to storage, whether on the surface or deep underground, remain complex issues. How to "conserve" these materials? Which cycles should they enter into? Part of irradiated materials generates time that exceeds the geological time scale.[10] The treatment of used nuclear fuel thus consists of separating

---

[7] It's APEC. See the diagram of this installation in *Superphénix, op.cit.* p. 56.
[8] Carbonation: neutralization of sodium by transforming it into sodium carbonate.
[9] It is managed by ANDRA (French National Agency for Radioactive Waste Management).
[10] The radioactive half-life of uranium 238 is 4.51 billion years.

Fig. 5.3 Detail of the Superphénix reactor floor during dismantling, seen from above. Copyright Jacqueline Salmon. All rights reserved

the materials (uranium and plutonium) which can be recycled in new fuel assemblies, and the so-called ultimate waste which is vitrified and poured into stainless steel containers (HAVL waste: "high long-lived activity").[11]

A nuclear power plant is in itself a memory. The dismantling of the installations is a "relentless revealer of the history of the power plant, and of more or less good operating practices; in terms of radiation protection and safety of work, particular attention must be paid to the unforeseen situations that we sometimes find".[12] As I saw, the power plant sometimes suffered from amnesia. Where were the plans for the plant, and who built it? It took some time before I managed to visit these sources. "Safety" is based on this work of archiving and memory necessary for later generations. A "return to nature" is not possible after the dismantling of the nuclear power plant.

5.5 Conclusion

The movement to rediscover the Kanak puzzle, like that of the rehabilitation of museographic objects, illustrates the skills for sharing and renewal which drive knowledge institutions. Inherited objects are a potential source of future wealth, backed by good conservation conditions. New growth always seems possible, and

---

[11] Operation carried out in the ORANO factory, La Hague, France (Documentation from the Creys-Malville site).

[12] Cit. Chapalain, see Bergé, Christine. Superphénix, des braises sous la cendre. *Monde Diplomatique*, April 2011, p. 5.

the notion of foundation remains at the heart of the process. However, time takes its toll on perishable materials, and the preservation of collections is always threatened. Restoration practices open controversies: it is also necessary to "document" the losses.

As for industrial objects, their temporality suffers from a recurring problem linked to the birth of the process: the process of extraction. Breaking the atom to extract heat results in the danger of exponential contamination (of workers and environment). The foundation of a nuclear power plant should therefore include the technical conditions for its closing-out. Certainly, the "experience feedback" process illustrates the communication shared in a precious collective memory. The problem of the "loss" of this know-how concerns the long term: the deconstruction of a nuclear power plant lasts decades. Non-knowledge also haunts the conditions of archiving ultimate waste, subject to non-human time. They threaten to be forgotten without having benefited from a full-fledged mourning, since they would be put in exile while waiting for the development of new know-how. This lack of knowledge causes concern since the burden will fall on the shoulders of later generations. Inheritance is therefore very problematic.

References

C. Bergé, *Superphenix, déconstruction d'un mythe*, Paris, La Découverte, collection Les Empêcheurs de penser en rond (2010). ISBN: 978-2359250374
C. Bergé, J. Salmon, *MHSD: Déconstruction* (Paris, Loco, 2011). ISBN: 978-2919507016
C. Bergé, J. Salmon, *Archives naturelles* (MARVAL Ed., Paris, 2002)
Coiffier, "Fougères et autres éléments végétaux associés aux casse-têtes kanak ou l'art de communiquer sans parole", in "La part d'immatériel dans la culture matérielle", Journal de la Société des Océanistes, p. 133–147 (2013). https://doi.org/10.4000/jso.6989

Open Access This chapter is licensed under the terms of the Creative Commons Attribution 4.0 International License (http://creativecommons.org/licenses/by/4.0/), which permits use, sharing, adaptation, distribution and reproduction in any medium or format, as long as you give appropriate credit to the original author(s) and the source, provide a link to the Creative Commons license and indicate if changes were made.

The images or other third party material in this chapter are included in the chapter's Creative Commons license, unless indicated otherwise in a credit line to the material. If material is not included in the chapter's Creative Commons license and your intended use is not permitted by statutory regulation or exceeds the permitted use, you will need to obtain permission directly from the copyright holder.

Chapter 6
Preserving and Valuing Memory for a Safer and More Sustainable Future: The Key Role of Archives

Fabienne Peris-Raimbault

Abstract The dismantling of industrial facilities and the closure of company sites are often an opportunity to rediscover little-known industrial heritage, especially archives, when they have not been destroyed. The preservation of this industrial, archival, tangible, or intangible heritage is often eluded because it is considered unnecessary and too expensive by companies. Taking aeronautics industry archives as an example, this chapter questions archives as a driving force for innovation, transmission of knowledge in a perspective of industrial safety and sustainability. Using the achievements of the past to build a more sustainable future!

Keywords Archives · Industrial heritage · Aeronautical heritage · Intangible heritage · Preservation · Industrial safety

6.1 Introduction

The dismantling of industrial facilities and the closure of company sites are often an opportunity to rediscover little-known industrial heritage and in particular archives, when they have not been destroyed. Indeed, not all companies are fortunate enough to have a qualified department, even if the function is developing in France (Orange, Safran, Saint-Gobain, etc.). Subject to numerous hazards, the archives may have been preserved by the heirs, the beneficiaries or the staff, constituted as an association for the preservation of cultural heritage, which is the case in the aeronautics industry. In the best case, these archives have been donated to a public archive service by way of

---

This work was supervised by Nathalie Regagnon, Deputy director of the Haute-Garonne Departmental Archives.

F. Peris-Raimbault (✉)
Haute-Garonne Departmental Archives, Toulouse, France
e-mail: fabienne.peris-raimbault@cd31.fr

donation or deposit, such as the French national archives of working life[1] (ANMT), a dedicated centre inaugurated in 1993.

Only public-sector companies are legally obliged to entrust their historical archives to a public archives service. In the private sector, "French tradition makes the state the driving force of awareness of the importance of archives and their organization" (Zuber 2006). And as a matter of fact, numerous recommendations, advice or training sessions are provided daily by public-sector archivists for private companies or other public institutions. But this cannot replace a professional archive department set up within the company itself. And yet… a company must keep all documents issued or received in carrying out professional activities for a minimum period of time, which varies depending on the nature of the documents and legal obligations. Civil, commercial, tax and social documents, accounting records, research and development projects, plans, know-how, oral testimonies and old archives are all resources that can be exploited by historians and the scientific community but also, and it must not be forgotten, by employees in their constant quest of innovation.

At a time of ecological transition, the preservation of this industrial, archival, tangible or intangible heritage is both a model and a source of inspiration for safety and sustainability. Indeed, thanks to rationalized practices of information management and conservation spaces, archive centres work in an ancestral way while respecting the environment.

Finally, by taking corporate archives as an example, in particular those of the aeronautics industry but also those of the administrations with which they are in contact, this chapter proposes to question archives as a driving force for innovation, transmission of knowledge and know-how with the aim of industrial safety. Using the achievements of the past to build a more safe and sustainable future!

---

[1] The National Archives of the World of Work (ANMT) are located in Roubaix, France https://archives-nationales-travail.culture.gouv.fr/

6.2 Multiple Preservation and Valorization Challenges

6.2.1 Managing Your Archives: A Legal Obligation and an Added Value for Your Heritage.[2]

Whatever its activity or its size, with a dedicated service or not, a company must keep a certain number of documents of economic and social character according to variable deadlines for legal, fiscal and organizational reasons. In the event of disputes with customers, suppliers or employees, it must be able to prove its commitments.

The archiving organization guarantees the durability of documents and rapid access to reliable and relevant information. Information management is a tool for strategic decision-making, which, in an often-competitive environment, is far from neutral (Hamon 2009).

Legal retention periods vary according to the nature of documents, but it is important to remember that archives can help to ensure the continuity of activities and serve as a reference for future action.

While the controlled collection and preservation of archives is an essential prerequisite for the legal and strategic protection of a company, it should not be forgotten that preserving content is not enough to keep its memory; it is also necessary to preserve its intelligibility, by documenting its production context and preserving the metadata necessary for its interpretation. At the opening of the AAF Archivists Forum in Troyes in 2016, Bruno Bachimont (lecturer and researcher at Compiègne University of Technology) mentioned the example of Concorde in 2008. He stated that when aircraft manufacturers wanted to produce a new civil supersonic airliner based on the Concorde that had been designed in the 1950s, the companies involved realized that they were unable to understand the Concorde design documentation, though barely 30 or 40 years had passed between the redesign analysis and the original design period. The companies had kept the documents but no longer knew how to read them.

Preserving the memory of activities is essential: the heritage function makes it possible to give meaning, but also to create links between employees, to reconnect with the company identity and to renew the image of the company.[3]

---

[2] Corporate archives and industrial heritage are recent concepts. At the end of the Second World War, the French state decided to create a Section of Economic Archives and a Section of Private Archives within the National Archives. Persuading company directors to protect and transmit their written heritage was one of the heavy tasks of Charles Braibant, then Director of the Archives de France. More generally, the preservation of industrial heritage is a relatively recent notion, dating back to the 1970s, especially since it is not the core business of industrialists, but in recent years, the concern has grown. The 2011 definition, and the Dublin principles by the International Committee for the Conservation of the Industrial Heritage, both take into account technical know-how, practices, etc. see: https://www.cilac.com/definition-histoire and https://ticcih.org/about/about-ticcih/dublin-principles/

[3] In Saint-Etienne, Groupe Casino deposited its collections and archives in the city's institutions from 1990 onwards. A scientific committee was set up in 2017 to ensure the relevance and scientific coherence of the work carried out and to coordinate the promotion of the collections. Several

But not all companies have worked on their history and, sometimes, are not ready to face up to certain dark moments dating back to periods of war, for example, "memory gaps" can be significant parts of a history to bury. However, taking a step back, clarifying certain events and being transparent can be an asset, but it is still necessary for a firm to possess its historical archives.

6.2.2 A Growing Heritage Function: The Example of the Aeronautics Industry

Even if the notion of aeronautical heritage was taken into consideration very early on by aviation circles, with the creation of a conservatory of aeronautics as early as 1919 in France (Aubagnac 2018), it was initially limited essentially to aircraft and other testimonies concerning the aerospace adventure. Over the last thirty years, this notion of heritage has expanded considerably.

In the case of the European aircraft manufacturer Airbus, whose origins date back to 1920 with the creation of Émile Dewoitine's manufacturing facilities, the term "aeronautical heritage" is used today to refer to the result of all the activities, from the design offices to the after-sales service, including aircraft, industrial buildings, production tools, intangible heritage (know-how and oral memory) and written and figurative heritage over a period of one century and on several sites in France. Such a protean concept (Lefebvre 2019) and such a wide temporal and territorial perimeter undoubtedly explain why the heritage notion was slow to develop, in the absence of an internal company policy common to its various sites in France and Europe. Until the creation of the Airbus Heritage department[4] in 2001, a dedicated service attached to the company's Communications department, conservation actions were mainly the result of personal or associative initiatives.[5] Thus, the vanguard role of associations has been paramount. Today, the *Airitage* association[6] federates all the heritage preservation associations located near the various French production sites in Toulouse, Nantes, Saint-Nazaire, Méaulte and Paris. If awareness of the heritage value of aeronautical assets, especially archives, is very present in the associative world, some items have completely disappeared and are strongly threatened (Olivier 2020). Nevertheless, the question arises of the treatment and valorization on a large scale of this heritage which is rich, complex, but often unknown, dispersed

---

exhibitions have been organized, « Vendre de tout, être partout. Casino» in 2020: https://mai.saint-etienne.fr/expositions-evenements/visites-virtuelles/vendre-de-tout-etre-partout-casino and the last in January 2024 « Casino, une histoire de famille», in line with the company's family history and to enhance the value of its employees: https://www.groupe-casino.fr/a-loccasion-de-leur-125-eme-anniversaire-les-enseignes-casino-mettent-alhonneur-les-femmes-et-les-hommes-qui-font-lentreprisea-travers-une-exposition-de-photographies-inedi/

[4] https://www.airbus.com/company/history.html

[5] For a summary of the issues surrounding Toulouse's aeronautical heritage, see (Péris-Raimbault 2021).

[6] https://airitage.fr/

and difficult to access. The current preservation activities of the Heritage Department of Airbus, strongly involved in the creation and operation of the Aeroscopia Aeronautical Museum,[7] are a useful contribution to this goal. How to make staff aware of the archival cause? Should everything be preserved? What conservation conditions? Can some document be scanned? So many considerations to which the *Haute-Garonne departmental archives*, with a specialized service in aeronautics, can bring its expertise.[8] In return, Airbus Heritage is contributing with its historical and technical knowledge of the company.

These different organizations in Toulouse continue their work today. This dynamic is positive but raises several questions about the process of heritage and sustainable conservation, which need to be addressed in order to continue with the objectives and professional working methods common to each of the entities. It can also create a common place of conservation and consultation commensurate with the challenge. Unless you have an internal dedicated service with a public website, only a structured public service guarantees the durability of documents and communication in compliance with the laws related to privacy, individuals' right to control the use of one's image, and intellectual and industrial property.

Heritage archives are an asset to ensure the cultural influence of a company. The industrial world too often forgets this, but they are also a source of inspiration in terms of innovation, training and risk management culture.

6.3 Burden or Inspiration? The Key Role of Archives

6.3.1 What Should Be Retained? What Should Be Eliminated? Focus on Archival Practices for a Safer and More Sustainable Future

Naturally, archive management is an inexhaustible source of inspiration for a safer and more sustainable future. The history of archives and archival practice is an ancient one, benefiting from feedback, rich in lessons learned at all levels. Whether collecting, evaluating and selecting, describing and indexing or conserving, reasoned and controlled working methods and methodologies can also provide sustainable solutions to risk management and major environmental issues.

The notion of information security and sustainability has always challenged archivists in several ways. You can't keep everything, and the archivist's first duty is… to destroy, in order to preserve what is essential! Preserving archives has a

---

[7] The opening of the Aeroscopia museum in Blagnac in 2015, dedicated to sharing the culture of aeronautics, bears witness to the desire to preserve and, above all, to promote a collection of works of art to the widest possible audience. https://www.aeroscopia.fr/

[8] https://archives.haute-garonne.fr/n/les-fonds-aeronautiques/n:137.

cost (building construction, server maintenance, etc.), among other things, the challenge is to reduce carbon footprint to a minimum. With volumes of information and documents increasing over the last few decades, especially in the context of digital massification, the evaluation and selection stage is an unavoidable, but complex, task for the archivist, which involves his or her responsibility. What should be retained? What should be eliminated? It's necessary to question the strategy to be adopted. Selecting what will allow future generations to write the history of our time implies a preliminary reflection that must be regularly reviewed (Rebours 2012), involving not only producers of documents but also researchers and users.

Gone are the days when the archivist sniffed files as if they were a melon and declared, usually without knowing why: this is interesting, this is not (Doom 2006). In France, a doctrine has only recently been formalized with the publication of the Methodological Framework for the Evaluation, Selection and Sampling of Public Archives, developed by the *Interministerial committee for French archives* (Juillet 2014).[9] This methodological tool can be used as a reference for company archivists because evaluation and selection are steps to be defined in advance with the employees and engineers who produce the documents. Evaluation is the archival operation aimed at determining the administrative, legal, historical and heritage interest of documents, ideally before they are produced. Depending on the company's industry sector, additional criteria may be defined: operational criteria linked to industrial safety, commercial criteria, scientific criteria (engineers may need to use the documents of their predecessors) and societal criteria. The choice of these criteria can be based on an analysis of the functions and processes performed or focus on the producers in order to understand their working environment, their missions and the documents produced, or to be produced. In the absence of procedures for processing activity documents, an organizational, functional or documentary evaluation survey may therefore be necessary.[10] In all cases, the aim is to control the production of probative and strategic documents and to avoid redundant documents being kept by different functional departments.[11] At the end of the assessment, the choice may be made for full conservation, partial or complete destruction.

Secondly, in the case of partial preservation, the selection stage aims to identify within the same set of documents, produced as part of the same function, which documents should be retained and those without interest which should be disposed of. For example, the retention of structured summary or synthesis documents such as registers, annual reports and statistics allows the elimination of part of the set they document at the end of the legal retention period.

---

[9] https://francearchives.gouv.fr/fr/file/5f01f41db3790b5201ff6c29413c16521a57ccf6/static_7742.pdf.

[10] See Appendix.

[11] Accounting documents will be kept by the accounts department only and will be destroyed in other departments.

In the fields of scientific, technical or industrial activities, evaluation grids, archiving tables and management of the useful life of documents can only be developed in consultation between experts and archivists or even historians and sociologists. Finally, beyond the legal framework of conservation, it has been, since the 1990s, important to draw attention to the value of personal scientific archives resulting from the research activities of company engineers or institutions or universities researchers. While final publications and books are well preserved, personal archives (laboratory notebooks, correspondence, minutes of meetings, reports, sheets, notes, field notebooks, various documentation) are often neglected (Lefebvre 2015). In the corporate world, the preservation of work archives is generally the result of personal initiatives such as André Turcat, pilot of the first flight of Concorde who donated its eight linear metres of documents to the *Haute-Garonne Departmental Archives*. By placing technological issues and advances in a broader strategic and diplomatic context, this source is proving invaluable in tracing the history of this unique aeronautical programme.

Another key stage is the archival analysis stage, which aims to organize the information content of a documentary collection in a precise and concise way, until the drafting of a research instrument, essential for sharing knowledge. Here again, archivists rationalize and optimize the management of information, prioritizing it for a quick and efficient access. All the guiding principles of description are defined by international standards.[12]

Sustainability, rationalization and economy are key concerns for archivists. Another aspect of the job directly linked to the ecological transition is resilience. Archives are evolving. Digital data is a new typology of data, in addition to traditional paper documents. Ensuring authenticity and securing access to data from all eras, in all forms and media, in the face of the proliferation of Fake News in the pressing context of the ecological transition, are issues which archivists tackle on a day-to-day basis.

6.3.2 Archives, Innovation, Transmission and Industrial Safety Culture

Some companies have turned away from their archives on the pretext that the profession they exercise today is not the one they exercised in the past, upset by the successive technological revolutions. In the race for innovation, the use of the past is not the priority, but it is not unusual to look for the anteriority of certain innovative developments in their time, or for a company to rely on its archives to continue innovating.

If the scientific world regularly relies on archive documents, the industrial world sometimes does the same. The use of the Airbus A300B at the Aeroscopia museum in

---

[12] ISAAR, ISAD(G) standards, which have evolved in particular to adapt to the semantic web (Records in context 1.0, published in 2023, is set to become an official ICA recommendation).

Blagnac (France) is exemplary. In the absence of the Flight test development aircraft[13] (dismantled in the 1970s) and without any digital mock-up, the Airbus engineers who provide customer support and maintenance regularly test the positioning of the latest generation equipment on this aircraft, which has also made it possible to develop and validate the dismantling processes for end-of-life aircraft.[14] In 2005, Airbus piloted the Process for Advanced Management of End-of-Life of Aircraft (PAMELA[15]) project. Today, the end of an aircraft's life is taken into account from the design stage.[16] Relying on the past, crossing memories sometimes saves time just like putting in perspective in the long run. Dismissing the past on the pretext that it would go against the principle of innovation is a prejudicial idea that must be questioned, particularly when the preservation of heritage is at stake.

The following examples illustrate the need to preserve the memory of industrial activities for training or safety reasons. Each year, the *Haute-Garonne departmental archives* receive a hundred requests, mainly by design offices, to access to the administrative and technical files of "Seveso" installations: these are often the only traces of polluting activities that have occurred at a given site. Having this information available upstream of any land development operation is essential: you can't build a school on a site that hosted a gas station for 30 years…

With the dismantling of old industrial sites, one is often faced with the loss of knowledge about old know-how or the use of machine tools. This problem is also old. The study of the exploitation of marble at Saint-Béat in the French Pyrenees, between the sixteenth century and the beginning of the twentieth century, has demonstrated that in the absence of continuous activity, the transmission of craftmanship and professional skills is lost. Under the Ancien Régime and in the nineteenth century, the French administration had to call on Italian porters from Carrara to lower the marble blocks without them being damaged. They were the only ones to master this leg technique thanks to the *lizzatura* method (Calestroupat 2020).

So, what about the memory of the craftmanship and professional skills? In some industrial sectors, the learning process for newcomers and the "operational experience feedback" process are a key issue. For a long time, know-how was passed on through companionship or mentoring. But now, because people are retiring, because human resource management is changing and because technologies are evolving

---

[13] The Flight test development aircraft ensures the reliability and maturity of the airframe, systems and powerplant for the Type certification campaign. After the entry-in-service, Airbus keeps one flight test aircraft and the digital mock-up to improve, test and certify all new equipment that is offered to or requested by the airline.

[14] https://documentation.musee-aeroscopia.fr/index.php/Detail/objects/402.

[15] Airbus and its partners (SITA France, EADS CCR, EADS SOGERMA SERVICES, La Préfecture des Hautes-Pyrénées), tested processes for deconstruction and recycling of parts or other materials from end-of-life aircraft, allowing operations to be carried out in safe and environmentally friendly conditions.

Today, the TARMAC AEROSAVE company dismantles end-of-life aircraft on its Tarbes site: https://www.tarmacaerosave.aero/

[16] https://aircraft.airbus.com/en/newsroom/news/2022-11-end-of-life-reusing-recycling-rethinking.

rapidly, we need to look at how knowledge can be preserved and passed on in formats that are easily and sustainably accessible. This is becoming a key issue when industrial safety and risk management are long-term concerns. In order to improve the transmission of know-how and knowledge related to craftmanship, the Research and Development department of French nuclear operator EDF has launched an extensive research programme, in the early 2000. This study focused on the optimization of video capture systems of specific work situations and tradecraft gestures in their context (Le Bellu 2010). This type of reflexive approach involving industrial techniques, cognitive sciences, information and communication technologies and archiving is costly and unfortunately too often ignored.

The Archives and Cultural Heritage service of the Haute-Garonne Departmental council is also involved in a research project to promote Haut-Garonnais industrial heritage. The study of the former sites of the paper company Riz-Lacroix serves as a starting point. Between 2019 and 2021, they also conducted a collection of audiovisual testimonies from former Riz-Lacroix workers. However, the cessation of activity and the loss of workspaces and machines did not allow to capture tradecraft, smells or noises. Faced with the problem of preserving professional skills and tradecraft, the public archives services are engaged in a race against time, as it becomes more and more difficult, if not impossible, to find former employees.

6.4 Conclusion

Far from being a burden, the preservation of historical documents is an essential memory tool that supports strategic decision-making, transmission of knowledge, technological innovation and communication, offering added value to the company. Thanks to archives, we can learn also from a safety point of view.

The profession of archivist is a strategic one, as it is the only guarantor of the preservation of the memory of current and past human activities. The challenge lies in rapid access to information, which can only be guaranteed by a competent and organized archive service, whether internal or external to a company.

Providing trustworthy documents and data over time is a daily challenge that requires resilience in its exercise. The profession is constantly evolving according to technological advances in information and communication. And in order to understand the complex missions and techniques of the industrial environment, collaboration between specialists and archivists is not only essential, but also indispensable as early as possible in the implementation of document management processes.[17]

---

[17] The example of the Archives Unit of the Nuclear Research Centre is representative of the good practices involving archivists and specialists: Lamare, Frédérick. 2016. La collecte des archives scientifiques d'un centre de recherche nucléaire: l'expérience de la cellule archives de Marcoule. *La Gazette des archives*: 145–155.
www.persee.fr/doc/gazar_0016-5522_2016_num_243_3_5387

Appendix—Some Criteria for Evaluation And Selection of Archives

In order to identify the documents to be kept and those without administrative or legal, historical or scientific interest and which can therefore be eliminated, here are some of the questions to ask:

An approach by producer of documents	
Identification	Name of the person Function and missions (objectives)
Context of document production	**What function in the company, what department?** *E.g. engineering, finance and accounting, HR, sales, manufacturing, customer service* **History of the function**, successive operating services and related functions **Political and/or societal interest**: public sensitivity to activities carried out within the framework of this function
Hierarchical positioning	What is the level of intervention *(e.g. supranational, national, local)*? If there are several producers, what are the functional links between them?
Evaluation of documents	**Volumetry ?** State of conservation, health status? Does a copy exist? Sustainability of document support? Are there summary documents? *E.g.: annual reports, activity reports, summary statistics*, etc
Review of information content	**Value of information content: primary/ secondary?** • Does the content of the documents have a social interest (genealogy, proof, law)? • interest in scientific research, for an industrial process or an innovation? • Does it have a specific historiographical need? • Does it have an aesthetic, emotional and/or symbolic quality? **Quality of information?** • Quality information is: significant, relevant, accessible, understandable, complete • According to the employee, what are the important files? **Status of documents ?** - Work documents? - Validated? - published? - Originals/copies?

(continued)

(continued)

An approach by producer of documents	
Risk review	• Legal risk (frequency and cost of litigation) • Physical risk (risk of contamination by infested documents, dangerous collection conditions, etc.) • Societal risk of not being able to satisfy public demand in the case of elimination

References

G. Aubagnac, P. Smith, (2018), Patrimoines de l'aéronautique, *In Situ*, 35: https://journals.opened ition.org/insitu/17008

S. Le Bellu, L. Saadi, 2010, Comment capter le savoir incorporé dans un geste métier du point de vue de l'opérateur? *ISDM: Informations, Savoirs, Décisions, Médiations*, 36 https://eprints.lse.ac.uk/33176/

A. Calestroupat, (2020), *Du Roi à l'Industrie : le règne minéral des marbres de Saint-Béat (XVIe siècle—début du XXe siècle)*, research memoire in history, supervised by E. Charpentier and P. Julien

V. Doom, L'évaluation scientifique des archives, principes et stratégies. Du melon au diamant. *La Gazette des archives*: 5–43 (2006). https://doi.org/10.3406/gazar.2006.3815

M. Hamon, 2009, Les archives de l'entreprise : actif matériel et gisement de ressources. *La Gazette des archives*:17–27. www.persee.fr/doc/gazar_0016-5522_2009_num_213_1_4527

M. Lefebvre, Projet PASTEL: PAtrimoine Scientifique Toulousain et Environnement Local. Research report halshs-02066215, University of Toulouse (2019)

Juillet, 2014, *Délégation Interministérielle aux Archives de France. Cadre méthodologique pour l'évaluation, la sélection et l'échantillonnage des archives publiques*. https://francearchives.gouv.fr/fr/file/5f01f41db3790b5201ff6c29413c16521a57ccf6/static_7742.pdf

M. Lefebvre, Projet PASTEL: PAtrimoine Scientifique Toulousain et Environnement Local. Research report halshs-02066215, University of Toulouse (2019)

J-M. Olivier, 2020, Machines et outils de l'industrie aéronautique. Un patrimoine menacé, à Toulouse comme à Seattle. *Patrimoines du Sud*, 11: https://journals.openedition.org/pds/3699

F. Péris-Raimbault, 2021, De Dewoitine à Airbus : vers une reconnaissance du patrimoine historique de l'aéronautique toulousaine, *Nacelles*, 11: https://interfas.univ-tlse2.fr/nacelles/1531

M. Rebours, S. Roussel, Vers une révision des critères de sélection des archives contemporaines : point d'étape du groupe de travail « évaluation et sélection des archives ». *La Gazette des archives*: 73–80 (2012). https://doi.org/10.3406/gazar.2012.4964

H. Zuber, R. Nougaret, Les Archives d'entreprises en France, *La Gazette Archives* (2006)

Open Access This chapter is licensed under the terms of the Creative Commons Attribution 4.0 International License (http://creativecommons.org/licenses/by/4.0/), which permits use, sharing, adaptation, distribution and reproduction in any medium or format, as long as you give appropriate credit to the original author(s) and the source, provide a link to the Creative Commons license and indicate if changes were made.

The images or other third party material in this chapter are included in the chapter's Creative Commons license, unless indicated otherwise in a credit line to the material. If material is not included in the chapter's Creative Commons license and your intended use is not permitted by statutory regulation or exceeds the permitted use, you will need to obtain permission directly from the copyright holder.

Chapter 7
Collecting, Dismantling, Documenting, Reusing: Marginal Practices with Discarded Electronics

Nicolas Nova, Anaïs Bloch, and Thibault Le Page

Abstract This chapter addresses how marginal communities of practice contribute to the dismantling of the large technical system made of obsolete electronic and digital objects. It relies on a field research investigation focused on practitioners in museums archiving computers and video game consoles, video game designers interested in vintage machines, artists who repurpose and reuse technological scraps in their work, as well as small organizations aiming to put unused devices back into circulation for social justice or education. Adopting the format of a visual essay, this contribution highlights how deconstruction happens in the ordinary everyday life of our digital society, and how these marginal practitioners produce various forms of knowledge and know-how about discarded electronics which are generally untapped.

Keywords E-waste · Discarded electronics · Reuse · Consumer goods · Margin

7.1 Introduction

The information and communication technology sector is characterized by rapid change and short obsolescence cycles, which affect both equipment and the know-how needed by practitioners. Devices and their software, identified by more or less distinct system versions (iPhone 15, PlayStation 5, Windows 11), as well as the infrastructures that enable them to function (as in the case of the 3G, 4G, 5G standards),[1] are all frequently updated and renewed. In other words, the IT sector is subject to a regime of constant innovation (Neff and Stark, 2003), which results in high levels of

---

[1] See for instance the website https://endoflife.date/, which tracks and documents "end of life" dates and support lifecycles for various digital products and services.

N. Nova (✉) · A. Bloch · T. Le Page
HEAD, Geneva, Switzerland
e-mail: anais.bloch@hesge.ch

T. Le Page
e-mail: thibault.lepage@hes-so.ch

© The Author(s) 2025
M. Bourrier (ed.), *Decommissioning Aging Installations and Declining Technologies*,
SpringerBriefs in Safety Management, https://doi.org/10.1007/978-3-031-88369-9_7

replacement for devices such as cell phones, televisions and video game consoles. Although the average lifespan of these devices depends on their nature—2–3 years for a telephone as shown on Fig. 7.1, 5 years for a games console—these figures are still extremely low compared to other technical objects, and especially to other industrial sectors. Since their proliferation in the 1980s, these electronic and digital objects have taken on a considerable role in contemporary consumer goods.

Their constant renewal means that these objects are only "new" for a very limited amount of time, after which their monetary or symbolic value diminishes considerably, even though they generally still function. This situation is a striking example of technologies that are permanently unstable. Even if this large technical system is not that of an industrial plant or site, it forms an interesting material and geographical assemblage, of which computers, smartphones and consoles are just the tip

Fig. 7.1 Two examples of machine and software storage collected by two associations, one in Paris (MO5, image on the left) and the other in Tokyo (Game Preservation Society (GPS), image on the right). In both cases, the objects accumulated over a period of twenty years are placed in labelled boxes, or sometimes on shelves. While in the Japanese case, this storage is carried out in the homes of the various GPS members (resulting in the distributed nature and sometimes messy of the collection), in the French case the objects reside in a warehouse. One of the most striking aspects of the practices investigated in our survey concerns the coupling between the collection of these "digital leftovers", and their storage. Accumulation requires a certain number of precautions to preserve these devices from material degradation. Hence, the presence of boxes to prevent dust, the use in the Japanese case of sensors (hygrometry) to monitor environmental conditions, the use of anti-fungal products, or the importance of removing batteries (informants note that old batteries can explode, leak, release harmful gases, or catch fire). Nevertheless, the use of boxes does not prevent the rapid accessibility of all these artefacts, which are constantly accessed by the members of these two associations, either with a view to organizing exhibitions (museums, festivals), for media shoots, or simply for the pleasure of this community of practice who reuse them from time to time. Unlike domestic storage arrangements such as attics or self-storage warehouses, which are generally described by sociologists as spaces of relegation before abandonment or destruction (Beldjerd and Tabois 2014), these two examples attest more to a logic of potential reactivation. Image credit: Nicolas Nova

of the iceberg. In order to function, all these devices are connected to each other by computing (data centres) and communications infrastructures (submarine or underground cables, antennas, satellites). One of the characteristics of this gigantic ensemble of mass-produced consumer goods is that it constantly produces obsolescence and waste. This makes it a particularly relevant case to explore, given its unstable and distributed nature.

This peculiar situation gives rise to a proliferation of obsolete objects, unusable applications and infrastructures to be replaced. There are many ways to describe this phenomenon: "electronic waste" (e-waste) or "waste electrical and electronic equipment" (WEEE), used electrical and electronic equipment (UEEE), or end-of-life electronics (EOL). While these terms are not equivalent, each has a negative connotation, underlining the loss of value of these devices, and their status between waste and scrap. Journalist and science-fiction writer Bruce Sterling refers to this as "Dead Media" and has proposed the creation of an archive documenting this graveyard of machines to map the forgotten, out-of-use and so-called obsolete technologies in order better to grasp the evolution of digital systems (Sterling 2015). Similarly, researchers Hertz and Jussi (2011) have coined the term "Zombie Media" to suggest that these obsolete and out-of-use objects have a persisting influence: they keep existing in ways that are different from what they were designed for, either by being diverted from their intended use, or by polluting the soil in landfill sites often located in Global South countries.

According to the United Nations Institute for Training and Research (UNITAR), 53.6 million tonnes (Mt) of e-waste were generated worldwide in 2019, which makes it the fastest growing waste stream worldwide, and among the top three categories of illicitly traded waste (UNITAR 2022). A threat to the environment and our health through the pollution they cause, the growing weight of digital waste is also a gigantic waste of critical metal resources.

In response to this situation, the manufacturers that produce such items adopt a variety of strategies, ranging from in-house recuperation and recycling to outright outsourcing, usually to countries in the Global South (Lepawsky 2018). However, a large proportion of discarded electronics never take such paths. Not least because these "digital leftovers", these obsolete or out-of-use technical objects that remain when their use is over or outdated, consequently end up discarded, sometimes at their owners' premises, sometimes in the trash mixed with other kinds of waste, or in more or less formal recycling circuits (Gabrys 2013). The said UNITAR report showed that only 17 of this volume was managed in an environmentally appropriate manner. What happens to the remaining 83 per cent is unknown, potentially recycled in an undocumented way, burned or dumped.

7.2 Observing "Under the Radar" Reuse

Alongside the processes adopted by manufacturers, which are legally required in certain countries, these obsolete technologies are also taken care of by multiple marginal communities, made up of designers, artists, tinkerers and, more broadly, people interested in upcycling,[2] reusing, and repurposing material goods. In this context, the term "marginal" refers both to the fact that these practitioners operate at the margin of industrial systems, and that the workforce they represent is limited. We will be looking at their activities to understand how they deal with electronic waste, unlike industrial installations and more standard waste management. The aim of our contribution is to show how marginal practitioners contribute to the dismantling of electronics. Without arguing that their actions can substitute for those of industrial groups, or for further regulation of this sector, we will rather show how they (1) contribute to a form of dismantling that occurs "under the radar" in the ordinary everyday life of our digital society, (2) they produce various forms of knowledge and know-how about them which are generally untapped.

To this end, we will draw on an ongoing field research of four communities of practice active in the collection and reuse of digital junk: (1) Digital preservation practitioners in museums and non-profit organizations archiving computers and video game consoles, (2) video game designers interested in vintage machines and "retrogaming" (collecting and playing outmoded or discontinued personal computers, consoles, and video games) or "chiptune" (music played with vintage computers and old consoles) practitioners, (3) artists in the field of *New Media Art*[3] who repurpose and reuse technological scraps in their work, (4) organizations and companies aiming to put unused machines back into circulation for social justice, educational or integration purposes. Since April 2022, our investigation of such practices has consisted of a series of online and face-to-face interviews, repeated observation sessions both in situ (in their workshops, during training sessions) and online (discord groups, forums, dedicated sites on which members exchange, produce, and share documentation that is produced individually or collectively). Adopting the approach of a visual essay commenting on a series of ethnographic vignettes, this chapter describes the repertoire of intervention we have identified in our fieldwork (Figs. 7.2, 7.3, 7.5 and 7.6).

---

[2] Also known as creative reuse, upcycling is the process of transforming obsolete materials, by-products, useless, or unwanted products into new materials or products perceived to be of greater quality, with an increased environmental or artistic value.

[3] New Media Art is an artistic field that encompasses focused on artworks designed and produced by means of electronic media technologies such as computer graphics and animation, virtual reality, networked communication, sound art, or any interactive systems.

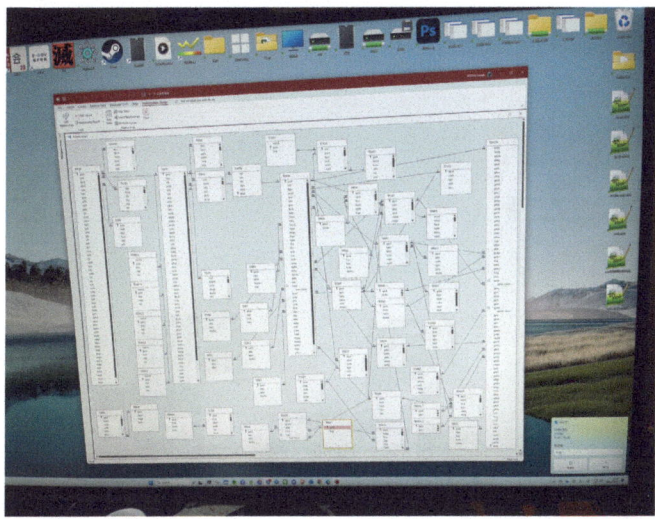

Fig. 7.2 Relational diagram of a database set up by one of the GPS members (Tokyo). This type of diagram represents the relationships between items in the collection assembled by members (computers, consoles, spare parts, video games running on some of these machines, documents about these items) as well as their location. Constantly updated and corrected by the members, this information system provides a detailed overview of the entire collection. The relational character of the database also allows to navigate and revisit the material preserved following a different logic than browsing shelves or picking items based on their shape or aesthetic. It also illustrates the important effort made to document these devices, which includes the production of a "duplicate" in a digital version that can be consulted to reactivate the use of the multiple elements brought together by this non-profit organization. Image credit: Thibault Le Page

7.3 Conclusion

The field vignettes presented in this chapter shed some light on how deconstruction happens in the ordinary everyday life of our digital society. More specifically, they show the role of marginal practitioners in dismantling a very specific kind of large technical systems made of information and communication technologies. The various practices we described here demonstrate how these communities of practice deal with electronic waste by collecting scraps, organizing repositories of discarded devices, reusing them to produce new artefacts. Depending on collectives, some do it by themselves or by putting together open workshops, repair cafés or training programmes, and sometimes generate various forms of knowledge with various forms of publications and art/design exhibits. While their local public is generally limited to dozens of like-minded colleagues or friends, the way they reach hundreds of contacts online through blogs, newsletters, video tutorials or reports make them important from a more global standpoint .

Fig. 7.3 Workshop during a yearly summer school in Nevers (France) in which participants are taught how to dismantle, tear-down, and repurpose spare parts of various electronic devices. After explaining the method and objectives, the team of workshop leaders invites participants to take apart various artefacts brought out from a series of suitcases represented on that illustration. Said participants are asked to do so either by removing screws and protective covers, or by damaging the device when this is not possible. If in this context the aim is to teach people to reuse whole devices or to pick up reusable spare parts, the intention of the organizers of similar workshops is often to overcome the fear of intervening on machines or simply the fear of dismantling objects. In other activities, the aim may be to stimulate reflection on our ability, or inability, to fix things, as in the case of repair cafés. Or it may involve an exploratory dimension, inviting participants to investigate the type of elements present inside these devices. Participants are then invited to open the objects, observe the electronic components and note the textual inscriptions present on them, before carrying out online research to identify their origin, role, and potential for reuse. Similarly, this kind of spontaneous exploration can also involve the content present on these old machines, by investigating the digital content present on discarded computers ("data forensics"); in this case for an artistic use aimed, for example, at alerting people to the issues involved in the dissemination of personal data

Without saying that these practitioners can replace state and commercial actors, or that the lessons to be learned from these cases are easily transposable to more traditional industrial infrastructures (nuclear power plants, factories), the presentation of these different vignettes illustrates a different dimension of dismantling linked to a very specific industrial system (decentralized, with hardware/software coupling, based on permanent renewal). The practices described in these pages nonetheless demonstrate the singularity of "under the radar" dismantling operations, based on a very specific relationship to these technical objects, but also turned towards a form of collective action.[4] This is based in part on the intrinsic motivation and very specific

---

[4] To some extent these "under the radar" practices could be compared to the production of factory "homers", e.g. artefacts that workers produce at the workplace and during workhours using company

7 Collecting, Dismantling, Documenting, Reusing: Marginal Practices … 79

Fig. 7.4 Fanzine[5] made by new media artist Garnet Hertz about how to use old battery powered toys to make a custom musical instrument or noise maker. Presented in the very rough aesthetic typical of fanzines, and distributed online via a Creative Commons open license, this 44-page document meticulously describes both in English and Spanish the procedure to put in place to complete such a project (This extract from the document *Toy Hacking Handbook: An Electronics Crash Course*, February 2015 Edition, published at https://conceptlab.com/circuitbending/, released under a CC BY 4.0 license.). This type of document is typically used in informal workshops or such as those shown in Fig. 7.4. It circulates online in digitized form, either free of charge or sometimes sold for a few dollars on craft sales sites (Etsy) or specialized community platforms (Reddit threads). As a counterpoint to official electronics manuals, these fanzines help to demystify these practices. They attest as much to the development of an expertise parallel to that of official training organizations as to a cruder logic of hijacking objects that are generally off-putting to non-specialists. Even if circuit-bending practices seem anecdotal at first glance, it is on the basis of these documents circulating online, or at workshops and seminars organized by artists or players in DIY culture that the practice of upcycling or repurposing material goods is developing. They can be seen as a complement to object disassembly activities, both because their content sheds light on the inside of the devices or their functioning, but also as an invitation to document these tear-down activities for oneself. Just like this "toy hacking" guide, or the many videos shared on online platforms, a multitude of similar documents offer a "grassroots" version of the technical drawings of the devices in question, which are generally not made available by companies. Basically, these are unofficial—and sometimes illegal—versions of information that technology companies do not wish to publicize, but which these activities reconstitute on their own[6]

---

tools and materials outside normal production plans (Anteby, 2006). Theirs is however different in the sense that they operate outside companies.

[5] A portmanteau word that comes from the combination of the words "fan" and "magazine", a fanzine is a paper-based publication produced by enthusiasts of a particular cultural phenomenon (such as a musical genre, underground topics or technologies) for the pleasure of others who share their interest.

[6] In recent decades, however, independent companies have been carrying out similar operations, offering for sale "tear-down reports" describing the results of their investigations; for example, reconstructing all the spare parts and circuitry of the latest smartphone models.

Fig. 7.5 Page spread of a book entitled "ENIAROF", which means "foraine" written backwards ("foraine" means "funfair" in French). Interested in revisiting funfairs, the authors published this book as a DIY guide to produce one's own attractions with video games, artistic installations and performances. A large chunk of the book focuses on procedures for subverting and modifying existing electronic devices as well as non-digital artefacts. Some of the proposals concern slight alterations, which hardly change the way the device works, while others are more radical, proposing ways of completely hijacking the functions of these devices. While not entirely focused on reuse, this book introduces in its own way to a wider audience what it means concretely to reuse and subvert existing devices. Photograph used with permission from the publisher; all rights reserved

relationship that the members of these communities construct in relation to technical objects, which they may value as much out of technical curiosity as nostalgia, or to derive a potential for misappropriation in design or artistic practices, notably to distinguish themselves from the existing. The importance of documentation, whether in the form of an information system to provide an overview of the objects collected, or via more or less formalized publications, also demonstrates the interest they have in the production of knowledge and know-how and in the sharing of these practices.

Without affirming that these generally untaped practices should be used or easily transposed to other fields either, they are in our view a relevant blind spot the industrial world should consider. There are three reasons for this:

- Dismantling is not depreciated, or seen as "a sad act of deconstruction" (the terminology used in the call for contributions of this workshop and publication). Rather, it is seen as a means of revaluing different elements of a technical system for what they actually are (computers, consoles and video games), or as a means of accessing reservoirs of spare parts that can be reused for other projects; and eventually fight against over-consumption. To some extent, some of the practitioners

Fig. 7.6 "Cubicle 2" by design studio FORMANTASMA (2017) was made out of CNC milled and folded aluminium and stainless steel, and the aluminum outer casing of a portable computer. This furniture was created in the context of an R&D project about electronic waste called Ore Streams (See their website for project documentation: http://www.orestreams.com/). Relying on such approach, these designers aimed at showing how rare earth metals (such as aluminum or gold) were increasingly found in other places than underground mines, enabling them to scavenge mobile-phone scrap and recycled metal in order to produce furniture. They call this approach "above-ground mining", illustrating how metal reserves are increasingly not underground but circulating in products or as electronic waste

we describe here correspond to the "menders" who seek to limit their environmental footprint that Bruno Latour talks about, and whom he distinguishes from the "extractors" who persist in exploiting the planet (Latour 2021).

- The repertoire of intervention, both material (dismantling, reuse) and intellectual (documentation), is very different in terms of the technical and cognitive operations performed and shared by these actors, with different nuances to the industrial forms of dismantling. Operating at a micro-scale and much more downstream, the activities undertaken by these practitioners eventually requalify what obsolete material and waste is.
- The operations carried out by these participants do not leave digital manufacturers indifferent, since they are sometimes interested in the many ways of rethinking the devices they produce to promote these activities. This corresponds to the notion of "design for hackability" (Galloway et al. 2004), the idea of designing artefacts and technical systems with their potential (re)appropriation in mind.

Beyond the field of information and communication technologies that concerns them, each of these points provides food for thought on the subject of dismantling. They show that, on the one hand, deconstruction and decay can nourish creative and motivated design and reuse activities. And on the other hand, the design of future objects and infrastructures should take into account their future reuse, and the ways in which they can be used in functions that may or may not be distinct from those for which they were intended.

Acknowledgements The authors would like to thank the Swiss National Science Foundation for funding the project on which this chapter is based (project No 204686).

References

A. Galloway, J. Brucker-Cohen, L. Gaye, E. Goodman, D. Hill, Design for hackability, in *Proceedings of the 2004 Conference on Designing Interactive Systems: Processes, Practices, Methods, and Techniques (DIS)* (ACM Press, New York, USA, 2004), pp. 363–366

J. Gabrys, *Digital Rubbish: a natural history of electronics*, (University of Michigan Press, Ann Arbor, 2013)

G.D. Hertz, J. Parikka, Five principles of Zombie Media, in *Conference: Defunct/Refunct* (2011)

B. Latour, *After Lockdown: A Metamorphosis* (Polity, Cambridge, 2021)

J. Lepawsky, *Reassembling rubbish* (Cambridge, MIT Press, Worlding Electronic Waste, 2018)

G. Neff, D. Stark, Permanently Beta: responsive organization in the internet era, in *Society Online: The Internet in Context* ed. by P. Howard, S. Jones, (dir.) (Sage Publications, Thousand Oaks), pp. 173–188

B. Sofian, S. Tabois, Le grenier, espace de retournement des choses. Socio-Anthropologie **30**, 21–31 (2014)

Sterling B, (2015), Dead Media Project. Available at the following URL: http://www.deadmedia.org/

UNITAR, (2022), Global Transboundary E-waste Flows Monitor—2022, United Nations Institute for Training and Research, Bonn, Germany). Available at ewastemonitor.info/wp-content/uploads/2022/06/Global-TBMwebversionjune2pages.pdf

Open Access This chapter is licensed under the terms of the Creative Commons Attribution 4.0 International License (http://creativecommons.org/licenses/by/4.0/), which permits use, sharing, adaptation, distribution and reproduction in any medium or format, as long as you give appropriate credit to the original author(s) and the source, provide a link to the Creative Commons license and indicate if changes were made.

The images or other third party material in this chapter are included in the chapter's Creative Commons license, unless indicated otherwise in a credit line to the material. If material is not included in the chapter's Creative Commons license and your intended use is not permitted by statutory regulation or exceeds the permitted use, you will need to obtain permission directly from the copyright holder.

Chapter 8
Safety Culture Lessons Learned in Decommissioning VTT's FiR 1 Research Reactor

Kaupo Viitanen, Merja Airola, Markus Airila, and Petri Kotiluoto

Abstract This chapter presents safety culture lessons learned from the decommissioning of the FiR 1 research reactor in Finland, the first nuclear facility to be decommissioned in the country. The authors reflect on their experience as leaders and safety culture experts in the decommissioning project and discuss the following four topics: the organizational and cultural aspects of the decommissioning strategy, the organizational identity, the changes in working practices and mindset, and the interfaces with external parties. The chapter provides practical recommendations and lessons learned for each topic.

Keywords Safety culture · Decommissioning · Nuclear industry · Research reactor

8.1 Introduction

The decommissioning phase of a nuclear facility is a *significantly different* context compared to its operating phase. It is characterized by constant change: the facility's configuration and risk profile change, and people, organizations, and the core task change. Decommissioning can be divided into sub-phases: pre-decision, post-decision operating, deactivation and defueling, storage period and dismantlement. As decommissioning progresses, nuclear safety hazards are eliminated, radiation hazards are generally reduced but change, and occupational safety hazards may increase. Hazards may also become more unexpected. Decommissioning often involves a network of (new) subcontractors, who may not be familiar with the nuclear facility in question, or with nuclear industry working practices. The business model

---

Disclaimer. The views expressed in this article remain the responsibility of the authors and do not necessarily reflect the opinions of other parties involved in the FiR 1 decommissioning project.

K. Viitanen (✉) · M. Airola · M. Airila · P. Kotiluoto
VTT Technical Research Center of Finland Ltd, Espoo, Finland
e-mail: Kaupo.Viitanen@vtt.fi

© The Author(s) 2025
M. Bourrier (ed.), *Decommissioning Aging Installations and Declining Technologies*, SpringerBriefs in Safety Management, https://doi.org/10.1007/978-3-031-88369-9_8

changes: decommissioning is a project with a beginning and an end, not a continuing production process like the operating phase.

A good *safety culture* is an established expectation in the nuclear industry. In Finland, legislation and regulatory framework require that each organization involved in designing, constructing, operating, and decommissioning of nuclear facilities shall maintain a good safety culture (STUK Regulation Y/1/2018). However, the existing safety culture and leadership knowledge base—the way in which safety culture has been conceptualized and concretized in the nuclear industry, and the commonly used models, practices, and methods—is largely shaped by incidents or accidents that occurred while operating nuclear power plants. Decommissioning involves different or new cultural challenges compared to the operating phase.

8.2 FiR 1 Decommissioning Project

Helsinki University of Technology and VTT Technical Research Centre of Finland operated the *FiR 1 TRIGA Mk II research reactor* in 1962–2015. The FiR 1 reactor has historical significance in Finnish nuclear research and industry. It was the first nuclear reactor in Finland and the only Finnish research reactor. FiR 1 was used, for example, for research, training nuclear power plant personnel, isotope production, and Boron Neutron Capture Therapy (BNCT) treatments of cancer patients. In 2011, VTT was granted a new operating license which would have been valid until the end of 2023. However, in 2012, VTT decided to shut down the reactor due to financial reasons. The decision to shut the reactor down so soon after receiving the operating license extension was faced with surprise and unenthusiastic feelings by VTT's staff and the whole Finnish nuclear community. The reactor closed permanently in 2015, and the decommissioning license granted in 2021 replaced the previous operating license.

FiR 1 is the *first nuclear facility to be decommissioned in Finland* (for further details, see Airila et al. 2022). A comprehensive contract on decommissioning services was signed in 2020 between VTT and Fortum Power and Heat (the main contractor). In 2021, the Finnish Government granted decommissioning license to VTT. All irradiated fuel from FiR 1 was delivered to the USA in 2021 to be used in a research reactor of the same type. As of the beginning of 2024, the reactor is being dismantled. The decommissioning project currently employs approximately 20 people at VTT, most of them part-time, and 20 people in the supply chain organizations. FiR 1 decommissioning is planned to be completed in 2024.

8.3 Safety Culture Lessons Learned

In this chapter, we elaborate *safety culture lessons learned* during the FiR 1 decommissioning project. The lessons learned are based on the authors' reflection and involvement in the decommissioning project as leaders and safety culture experts. The first author is an independent safety culture expert on the FiR 1 safety committee and has facilitated FiR 1 safety culture self-assessments since 2021. The second author is a safety culture expert who has been involved in all FiR 1 safety culture self-assessments since 2018. The third author is the decommissioning manager of the nuclear facility and the project manager of the decommissioning project. The fourth author is the responsible manager of the nuclear facility.

We focus on four essential safety culture phenomena we observed during the decommissioning project and examine the following: How were they identified? What was done to manage them? What could be learned based on our experience? What is the transferability of the lessons learned to other contexts?

8.3.1 Organizational and Cultural Aspects of the Decommissioning Strategy

Decommissioning strategy defines the timeline and end state of the decommissioning project. IAEA Safety Standards recognize two general categories of decommissioning strategies: immediate and deferred dismantling, with the former being preferable (IAEA 2023, 2014). In case of immediate dismantling, the decommissioning starts right away (as soon as technically and administratively feasible) after the permanent shutdown. In case of deferred dismantling, there is a period of safe enclosure when the site continues to be maintained. The decommissioning strategy is influenced by factors such as the extent and spread of contamination, the intended use of the site after decommissioning, etc. (IAEA 2023) In addition to these technical and administrative considerations, the decommissioning strategy also involves organizational and cultural dimensions, which we will focus on.

Immediate dismantling was chosen for FiR 1 decommissioning. The licensing process started with a lengthy environmental impact assessment (EIA, carried out during post-decision operation). The dismantling started in 2023, eight years after the permanent shutdown, because of the long licensing process (2017–2021) and various delays. One of the main reasons for the delays in this phase was nuclear waste management contracts which were not in place at the time the decommissioning license application was submitted. Overall, there were two time periods characterized by uncertainty that may have had an impact on the staff and the organizational culture: the operating period after the surprise decision to shut down the reactor (2012–2015), and the prolonged license review process. The post-decision operating period was faced with mixed feelings. On the one hand, the staff found it positive that the EIA process was long, which meant that the operation could continue for a few more years.

On the other hand, many perceived the decision as absurd, particularly considering the recent extension of the operating license.

There was uncertainty in terms of *work planning and orientation*. It was not clearly known when the next project steps would take place. This made it difficult for the personnel to prepare and for the managers to maintain the motivation of the personnel. Personal development discussions with personnel were an essential tool to identify and manage the uncertainties of the lifecycle phase transition. Still, the *root cause* behind these uncertainties was the *surprise shutdown decision*, which meant that the essential plans (incl. nuclear waste management contracts) were not available when the licensing phase for the decommissioning started.

Unlike many other decommissioning projects where operating personnel are replaced with decommissioning personnel, the VTT approach to manage the transition was to utilize almost all reactor *personnel from the operating phase* in the decommissioning project. There were many reasons behind this decision. The decommissioning phase shares many organizational functions with the operating phase, such as nuclear safeguards, security, radiation protection, and emergency preparedness. This means that there was a need for qualified personnel for these tasks and the operating personnel already had the necessary regulatory certifications. The operating personnel also had in-depth knowledge of the reactor and its history. The decision to use operating personnel in the decommissioning project had many advantages and disadvantages.

First, it meant that there was a need to *break away* from routinized working practices and mindsets, and learn new ones (see Sect. 8.3.3 on working practices and Sect. 8.3.4 on working with external parties). It was a significant change in culture, which was not always easy to implement because of the long operating history, typical long careers at VTT, and established routines.

Secondly, it meant that the personnel had formed a certain *attachment* to the reactor as an artefact of their working place, which affected the motivation and approach to the decommissioning project. For example, there were tendencies for extreme thoroughness and care in planning or preparation or trying to find reuse for the components of the facility, which sometimes adversely impacted the decommissioning progress.

Thirdly, the operating personnel—particularly those with long working experience at the plant—were an important *source of information*. The reactor operation dates to the 1960s, which means that the documentation culture, practices, and systems have varied a lot. A lot of information is tacit, unavailable, or difficult to find in documentation. The experienced personnel were able to bring their in-depth knowledge of the reactor to the decommissioning project—something that no-one else had. This was crucial when the decommissioning project faced problematic or uncertain situations. An example is the investigation of a water leakage from the reactor tank in connection with dismantling of the internal components of the tank. The experienced members of FiR 1 personnel were able to bring up and interpret essential photographs from the time of construction of the reactor. This information was used in the root cause analyses of the water leak. The possibility of still being able to contribute was highly motivating for the personnel.

Fourthly, taking the whole *operating organization* along to the decommissioning project means that the resulting organization may include people who are not essential in decommissioning. For VTT, it was important to maintain the feeling of job certainty and stability among the personnel and no conscious decisions were made regarding what type of people or competence profiles were needed for the decommissioning phase. This eventually resulted in the formation of "core personnel" and "peripheral personnel". The integration of the "peripheral" group into the project was a challenge because their technical competences were needed only to a limited extent during the prolonged planning and preparation phase (e.g. procedure development, and generally in leading and managing the project) while the "core" group was continuously fully occupied with administrative tasks. In hindsight, the formation of the two groups might have been avoided with more strict decision-making in terms of size and selection of staff, or with different strategies in ensuring that the "peripheral" group felt included and is involved.

Lessons learned:

- Identify the *organizational and cultural dimensions of the decommissioning strategy*. Consider who will be involved, what they will do, how their interest and motivation will be affected in the decommissioning phases, and what can be done to maintain and improve it.
- *Plan the decommissioning project* in sufficient detail (incl. contracts, spent fuel arrangements, waste management) and early in the overall process, but preserve enough flexibility for the unexpected. Having a timely and detailed overview of the decommissioning project helps lead the employees and prevent the feeling of uncertainty.
- Identify and acknowledge the advantages and disadvantages of using *personnel from the operating phase* in decommissioning. Recognize who are essential—from competence and experience perspectives—and make an effort to ensure their commitment and acculturation to the demands of the new life cycle phase.

8.3.2 Organizational Identity

VTT is an applied technological research institute. Its *organizational identity* values innovation and scientific discoveries, and its purpose is "bringing together people, business, science, and technology to solve the world's biggest challenges, creating sustainable growth, jobs and well-being". Indeed, implementing a pragmatic and straightforward decommissioning project was not considered interesting or motivating for VTT or for the employees who worked at the reactor. Instead, it was decided to treat it as a scientific research and development project—something that is more naturally in line with VTT's identity.

A strong *emphasis on research and development*, which incorporated a long-term vision into the decommissioning project, was a unique aspect of FiR 1 decommissioning strategy. The decommissioning project could have been implemented with lesser resources if the only goal was purely implementing the decommissioning.

Instead, the FiR 1 decommissioning was seen as an opportunity of learning for VTT and its partners (e.g. Finnish nuclear power companies, which will have to prepare for decommissioning their older plants within the next decades). This parallels the original purpose of the research reactor as the forerunner in the Finnish nuclear industry. To make this possible, a development project "dECOmm", funded by the Finnish innovation and R&D funding body Business Finland, was initiated. Within this project, VTT launched a decommissioning business ecosystem to develop new services for the international decommissioning market together with other Finnish companies where the FiR 1 decommissioning has served as a test bed for new decommissioning technologies. These included comprehensive computational characterization of radionuclide inventories, which resulted in several scientific publications and a doctoral thesis, and which was also instrumental for the planning of several technical aspects in the decommissioning project. Other topics in which VTT has been actively disseminating lessons learned are spent fuel transfer arrangements, licensing, radiation surveys, waste management, intercomparison exercises for difficult to measure radionuclides, and new sampling techniques.

These scientific and future-oriented initiatives were ways to facilitate employee commitment and motivation in the decommissioning phase and to create continuity in relation to the operating phase. Indeed, the decommissioning strategy which placed emphasis on research and development became a shared identity for VTT employees who worked on the project and something the employees are proud of. It also provided them with prospects as the accumulated expertise is widely valued and helps ensure future project portfolios (e.g. commercial and EU projects).

However, while the research and development orientation was a natural way for VTT to create motivation, commitment, and continuity, this might not be directly *transferable* to other decommissioning projects. For example, other decommissioning projects might not keep the operating personnel and might rely more extensively on outside staff, or they might not value research and development to the same extent and instead apply a more pragmatic decommissioning strategy. Potential problems may arise when the operational staff is kept, but the organization is not able to identify ways to ensure their commitment—something that we also found indications of in the form of "peripheral personnel". Research and development novelty value in decommissioning projects may also be reduced in future when they become more commonplace and routinized.

Lessons learned:

- Understand the existing *identity of your organization*. Examine what motivates people and what causes uncertainty. Leverage this information to drive the transition to the decommissioning phase.

8.3.3 Working Practices and Mindset in the Operating Phase and in Decommissioning

As the organization transitions from the operating phase to decommissioning, its *core task changes*. A change in core task triggers demands for changes in working practices and mindset.

In the operating phase, activities at the FiR 1 reactor were highly *routinized*. The risks, reactor characteristics and task demands were stable and known. Tasks of reactor operators and their supervisors (shift supervisor) were established, and procedure based. The operators controlled reactor power levels and performed the necessary checklists. During a typical day, the operator was actively required for one or two hours. The other activities at the reactor, such as isotope production or irradiations, were similarly routinized. The routines at the reactor had been created during the long history of FiR 1 operation and then passed on to the new generations, who did not need to "reinvent the wheel". The total number of operating procedures was relatively low. Continuous improvement at the reactor was mainly based on the use of operating experience, resulting in minor procedure updates, which were then disseminated and discussed during annual training days.

When the FiR 1 reactor went on to decommissioning, *work management practices* changed. The preparatory phase was organized as sprints, which resulted in a higher tempo in executing project tasks than typical in research. During the dismantling phase, VTT defined that the on-duty operator and shift supervisor are fully committed to overseeing the contractor's work at the site. At VTT, work is organized in parallel projects, which meant that (almost) every VTT employee involved in the decommissioning also worked on other projects that were ongoing at the same time. The sprint work was experienced as a positive development as it facilitated a constructive discussion about prioritizing work tasks. However, the full commitment of the shift personnel in the dismantling phase caused challenges for work planning and work time management, ultimately resulting in many people working under high stress and workload. The high stress and workload contributed to negative safety culture issues such as difficulty finding time for continuous learning and development, which are important tasks during the dynamic decommissioning phase. VTT acknowledged the workload issue throughout the project and tried to address it by assigning additional resources to specific areas and improving work shift arrangements, which somewhat improved the situation.

The FiR 1 decommissioning project required significant *planning* effort and *development of many new working procedures*. Even though many of the VTT employees participating in the decommissioning project were also involved in procedure development, this was not the case for everyone. This sometimes resulted in gaps in knowing all the procedures or not knowing the most up-to-date procedure, which was further accentuated by constant changes in the decommissioning situation. The high number of new and changing procedures caused a challenge for communication. In the complex and dynamic decommissioning phase, the previous practice of annual training days was not sufficient for informing about the new procedures and jointly discussing the procedures with the employees. Annual training days were supplemented by thematic training on key topics such as site working practices, radiation protection, and waste management. The procedure development work also set new demands for document management systems, which were not ideal for the decommissioning phase, and which were consequently restructured and centralized to one workspace.

The decommissioning project also marked the beginning of a wider development of the FiR 1 management system and *systematic improvement of safety culture*. This was partially triggered by the regulator's increased interest in FiR 1 as the organization underwent the transition from operating to decommissioning phase (see also Sect. 8.3.4). FiR 1 management has been implementing a systematic safety culture development plan in the decommissioning project since 2016, which included four safety culture self-assessments between 2018 and 2023 (see, e.g., Ylönen and Airola, 2018), safety culture trainings, inclusion of safety culture topics in project meetings, and implementing various organizational development actions based on assessments and other observations. There has also been an independent safety culture expert present in the FiR 1 safety committee since 2017.

Lessons learned:

- Acknowledge that *learning the new organizational culture*, working practices, and mindset when moving from operating to decommissioning phase *will take time*.
- *Leaders should understand and effectively communicate* what changes when the organization transitions to decommissioning and help create a shared understanding of the expected working practices.
- Ensure that the *preconditions are in place to reorient* people to the working practices and mindset. This includes assigning sufficient time and resources for learning, development, and communication, both for the staff, but also for the leaders.
- The lifecycle phase transition can *break organizational stagnation and complacency*, which is an opportunity for overall development. Identify these opportunities for development and try to benefit from them.

8.3.4 Interfaces with External Parties

Finnish society is characterized by high acceptance of nuclear technology and trust in regulatory oversight. Consequently, the surrounding society was not considered as a stakeholder that would require significant attention (or "active stakeholder management") in the FiR decommissioning project. Instead, here we focus on the regulator and supplier interfaces as the most important external parties.

The *regulator* was the primary external interface in the operating phase. Like the activities at the reactor, regulatory oversight had become routinized and implemented by inspectors with long experience in overseeing FiR 1. From a regulatory oversight perspective, FiR 1 operation was largely uneventful. For example, the regulator did not note significant nuclear or radiation safety events during the last decade of FiR 1 operation.[1] The lifecycle phase transition not only challenged the licensee but also the regulator: it was the first decommissioning project implemented in Finland and a major change in a previously "stagnant" nuclear facility. The regulator took a stricter and more active approach. They established a project for overseeing FiR 1 decommissioning and expanded their expertise base of inspectors involved in FiR 1 oversight. Safety culture, leadership, and organizational issues became a recurring topic in inspections. At the same time, two nuclear power plant newbuild projects were ongoing in Finland, and the regulator applied the experiences and lessons learned from their regulatory oversight to overseeing FiR 1 decommissioning. The open and proactive relationship between FiR 1 management and the regulator facilitated the development and systematization of FiR 1 processes and practices. Examples of improvements made during the lifecycle transition where regulatory oversight played a role included extensive FiR 1 management system improvement, implementation of a safety culture development plan, and development of procedures for supply chain management.

When the FiR 1 reactor went on to decommissioning phase, VTT's role changed from the only party at the reactor to an overseer of other parties. New external parties were introduced in the form of the *supply network*. Their contract covered dismantling and nuclear waste management services. Outsourcing was crucial because VTT did not have the necessary industrial capability to conduct these tasks on its own. The suppliers provided their industrial experience and well-proven processes in nuclear waste management and dismantling, and VTT provided scientific expertise, method development, and local knowledge of the reactor.

The transition towards dismantling demanded integrating the different organizational cultures, where key factors were close cooperation, open communication, and a healthy questioning attitude from all parties involved. Even though FiR 1 operating organization had no well-established *supply chain management processes*, FiR 1 management had assumed that it would be a relatively straightforward task to implement. The size and complexity of the supply network in the decommissioning phase was small, and the suppliers were local, familiar, and known to be highly competent. Comparing FiR 1 decommissioning to the complex and multicultural challenges of the recent newbuild projects in Finland, this assumption can be seen to be justified. Despite this, some challenges were still experienced.

One challenge related to the assumption of *cultural compatibility* between the parties. VTT's organizational culture and practices formed during the long operating phase of FiR 1 reactor and reflected that context, while the organizational cultures of the suppliers reflected their industrial experience in waste management and annual maintenance of nuclear power plants. Each organization's culture was "good" in terms of being aligned with the demands of their organizational core task, but they had different approaches and working practices to achieve it. VTT's personnel were familiar with the reactor and felt confident in working at the site due to their lengthy

---

[1] Source: Finnish Radiation and Nuclear Safety Authority's Annual Safety Oversight reports between 2000–2011.

experience. However, they were not always familiar with practices specific to decommissioning or with the practices agreed with the other parties. Sometimes there were also disagreements because the expected decommissioning phase practices were different than in the operating phase. These created tensions between the parties. Indeed, for VTT, one of the organizational culture development challenges was to come to terms with the idea of having other parties at the reactor. This meant adopting the new oversight role, breaking away from the old operating phase practices, and learning the new working practices.

Communication between the management and the operative personnel was essential in solving these issues. The operative personnel might not always know what or why something has been agreed with on the management level between the different parties. There was a tension between "work-as-imagined" (procedures agreed between the parties on a managerial level) and "work-as-done" (working practices inherited from the operating phase and new working practices specific to the decommissioning phase). A specific leadership challenge is effective communication of arrangements that have been made on a managerial level.

Safety culture self-assessments were one of the tools to proactively identify organizational and cultural areas for improvement. Topics and questionnaire items concerning the management of third parties were included in all our safety culture self-assessments since 2018. However, their focus was primarily on reflecting VTT's own supply chain management capacity and VTT's own perceptions of the supply chain. In 2023, during dismantling phase, the FiR 1 safety committee decided to also include supply chain personnel as a data source. This external perspective gave a much better understanding of the cultural dynamics between the different parties involved and initiated a constructive project-wide discussion which helped VTT better identify our cultural strengths and weaknesses and grow as a licensee.

Lessons learned:

- Acknowledge that *creating a shared understanding* of the project, its goals, working practices and mindset is essential in smaller projects with smaller supply networks too. Do not underestimate the effect of organizational culture differences between seemingly similar or compatible companies.
- Actively *involve all parties and all levels of the organizations* early, and during the entire decommissioning project. Make sure that leadership expectations and working practices are agreed and communicated effectively within and between all parties. Consider contractual arrangements that encourage taking joint responsibility for project implementation (e.g. project alliancing).
- Identify the *organizational cultural characteristics of the parties* and discuss their implications together. Prepare a joint plan to manage their risks and make an effort to ensure good collaboration.
- Interaction with other organizational cultures can help reveal cultural strengths and weaknesses and blind spots. Ensure *open and constructive communication between the parties* and facilitate a healthy and respectful *questioning attitude*.

- Include *external (supply chain) perspective in safety culture self-assessments* to help recognize potential issues in cooperation. Discuss the findings and their implications with all parties involved.

8.4 Conclusion

Decommissioning of nuclear reactors will be a significant challenge for the global nuclear industry as older facilities are nearing the end of their lifetime. The FiR 1 decommissioning project has been a unique learning opportunity for the Finnish nuclear industry, a driver for renewal, a testbed for method development, a source of new ways of working, and a case study for scientific research. In this chapter, we described some of the safety culture lessons learned from FiR 1 decommissioning. These lessons learned can be valuable for future decommissioning projects.

References

M. Airila, I. Auterinen, J. Helin, T. Kekki, P. Kivelä, P. Kotiluoto, A. Leskinen, A. Räty, Case Study on Decommissioning of the FiR 1 TRIGA reactor, VTT Research Report. VTT Technical Research Centre of Finland (2022)

IAEA, Decommissioning of Facilities (No. GSR Part 6). International Atomic Energy Agency, Vienna, Austria (2014)

IAEA, Global Status of Decommissioning of Nuclear Installations. International Atomic Energy Agency, Vienna, Austria (2023)

M. Ylönen, M. Airola, FiR 1—Tutkimusreaktorin käytöstäpoiston turvallisuuskulttuuriarviointi (No. VTT-R-0216-18). VTT Technical Research Centre of Finland, Espoo, Finland (2018)

Open Access This chapter is licensed under the terms of the Creative Commons Attribution 4.0 International License (http://creativecommons.org/licenses/by/4.0/), which permits use, sharing, adaptation, distribution and reproduction in any medium or format, as long as you give appropriate credit to the original author(s) and the source, provide a link to the Creative Commons license and indicate if changes were made.

The images or other third party material in this chapter are included in the chapter's Creative Commons license, unless indicated otherwise in a credit line to the material. If material is not included in the chapter's Creative Commons license and your intended use is not permitted by statutory regulation or exceeds the permitted use, you will need to obtain permission directly from the copyright holder.

Chapter 9
Decommissioning Management and Leadership for Safety Education: Addressing the Organizational Challenges and the Managerial Complexity of Nuclear Decommissioning Projects

Yoann Guntzburger, Jacques Repussard, Savéria Cecchi, Pierre Daniel, Renata Kaminska, Joseph A. Ridao Cabrerizo, Evelyne Rouby, and Catherine Thomas

Abstract Numerous countries are facing the phase-out of ageing nuclear installations, ranging from power reactors to medical and research facilities. There is already a wealth of technical, financial, regulatory, and societal knowledge amassed from past nuclear decommissioning and dismantling (D&D) efforts. Most of this collective experience is formalized in the publication of specialized guidance and information documents, particularly within the framework of the IAEA and the OECD/NEA. However, a critical analysis reveals a gap in the comprehensive examination of these challenges from a managerial perspective. This gap hinders the effective translation of theory into practice, challenging practitioners to steer complex D&D projects without a solid grasp of the socio-organizational dynamics at play. In this chapter, we present the Decommissioning Management and Leadership for Safety Education (DMaLSE) project, co-funded by the European Union. The primary objective of this project is to develop and deliver master-level training, tailored for professionals engaged in the management of nuclear D&D projects. By employing a systemic

Y. Guntzburger (✉) · R. Kaminska
SKEMA Business School, Université Côte d'Azur (GREDEG), Sophia Antipolis, France
e-mail: yoann.guntzburger@skema.edu

J. Repussard
Institut Pour La Maîtrise Des Risques, Cachan, France

S. Cecchi · E. Rouby · C. Thomas
Université Côte d'Azur, Nice, France

P. Daniel
SKEMA Business School, Lille, France

J. A. Ridao Cabrerizo
Karlsruhe Institute of Technology, Karlsruhe, Germany

© The Author(s) 2025
M. Bourrier (ed.), *Decommissioning Aging Installations and Declining Technologies*, SpringerBriefs in Safety Management, https://doi.org/10.1007/978-3-031-88369-9_9

approach to model the D&D process, conducting an integrative review of the literature, and incorporating expert opinions, we have pinpointed crucial managerial topics and their interdependencies. These elements form an analytical framework for examining various ongoing D&D cases. These case studies will be part of the teaching resources used during the forthcoming master-level professional training programme.

Keywords Nuclear decommissioning · Leadership for safety · Management · Education

9.1 Introduction

Until recently, the dismantling of nuclear installations (power plants, research reactors and laboratories, nuclear fuel industry, and medical facilities) was not factored in their design. This contributed to the many challenges facing nuclear decommissioning projects, some lasting several decades. For example, the decommissioning of the Brennilis nuclear power plant (NPP), one of France's first reactors, which started in the 1990s, is not expected to be finalized until 2042, after a long series of stop-and-go decisions and technical difficulties. In 2016, the French parliament set the target date of 2030 for the removal and safe definitive storage of historical nuclear waste accumulated on several French nuclear research and industry sites; however, this target will not be achieved, and clean-up operations are expected to last well into the 2040s. In the 2000s, because of unanticipated technical difficulties, the project of building and operating a specially designed installation for interim storage of spent nuclear fuel in the remaining three nuclear reactors of Chernobyl NPP, funded by the European EBRD, had to be abandoned halfway, and was eventually re-designed by a different contractor.[1]

Of course, there are also success stories in nuclear decommissioning projects, in France and elsewhere, such as that of EDF's first pressurized water reactor (PWR) at Chooz NPP, or Finland's research reactor at VTT[2] However, most countries with a history in nuclear research and industry are encountering similar difficulties in decommissioning processes, which tend to set nuclear dismantling projects aside from similar projects in other industries. The question arising is whether these challenges stem solely from the nascent state of nuclear decommissioning methods, compounded by the specifics of early nuclear technologies, or if they also arise from inherent difficulties specific to the nuclear sector. Subsequently, this raises the issue of the learning trajectory required to effectively navigate these particular challenges, especially as an increasing number of countries must address the decommissioning of their nuclear infrastructure.

---

[1] https://journaldelenergie.com/nucleaire/arevas-incredible-fiasco-in-chernobyl/

[2] See (Viitanen et al., this volume).

9.2 The Specific Nature of Nuclear Decommissioning Programmes

There are at least three specificities of nuclear decommissioning programmes:

- **A paradoxical relationship to passing time.** Contrary to stable chemical elements, radiation-related toxicity of nuclear isotopes declines over time, partly reducing the challenge associated with their manipulation during dismantling processes. In parallel, in some countries—Europe in particular—the law requires nuclear operators to provision and safeguard adequate funds dedicated to dismantling works during the operating phase. From a boardroom perspective, these funds grow in future value faster than the often-undervalued dismantling costs, giving a false impression that decommissioning will be easier and less expensive tomorrow than today… However, the detailed knowledge of inactive nuclear installations also declines with time, and because of this, many key stakeholders, including nuclear regulators, tend to favour an early initiation of decommissioning, as soon as possible after the end of the production phase.[3] These paradoxes and opposite injunctions are likely to create a context where the sense-making of daily work processes is elusive at operational levels.

- **The pervasive sociocultural burden associated with radioactivity and radiation exposure risk.** The nuclear industry is the only one where public acceptance is closely conditioned to the capacity to safely confine radioactive material at all times. This societal attitude, commonly shared in many countries at varying degrees, has several causes, mainly related to an oversized radiation risk perception fed by the collective memory of the atomic bombing in Japan and recent major nuclear accidents. This has some positive implications, such as the existence of a unique international nuclear safety and security framework at the United Nations level, through the IAEA, and of stringent national policies which include regulatory requirements for confining radioactive materials at all stages of their life and managing the safety of nuclear waste.

  As a result, the feasibility of nuclear decommissioning projects is de facto conditioned by the availability of radioactive waste disposal facilities. In turn, waste packaging and transport-specific requirements constrain dismantling operations, potentially conflicting with their own technical and operational rationale.

  Because decommissioning projects will involve operations, transport in particular, which may affect the potential exposure of local communities to radiation, each project's acceptance is also an issue in itself, requiring openness to stakeholders, and understanding of their perceptions and expectations. Failure to act on this may severely affect the project's course over time.

- **The confrontation with uncertainty, in an industry groomed to observe all-important rules and procedures derived from deterministic safety demonstrations and aimed at suppressing uncertainty.** Decommissioning projects often require temporary installations for specific operations such as the sorting

---

[3] IAEA safety Standards, decommissioning of facilities, General safety Requirements, Part 6.

and packaging of radioactive waste, or robot-operated procedures. Digital reproductions of parts of installations can also be used to validate operating plans and to train operators. It has been repeatedly observed that engineering groups with newbuild nuclear industry experience tend to underestimate the high uncertainty levels characterizing nuclear decommissioning projects, possibly leading to conceptual errors that could hamper the projects in several ways. Similarly, safety management approaches must consider that multilayer defence-in-depth protections built into the installation's design and operating procedures will be gradually removed or modified as the dismantling progresses, requiring a bird's eye view on safety issues, an ongoing critical review of protection procedures, and mindfulness at all times on the part of managers and operators during sensitive work phases. The dominant regulatory culture may well be out of step with such changes and uncertainty challenges, further complicating matters.

The nuclear industry and its regulators are mostly aware of the above problems, which have been (and continue to be) in evidence in many decommissioning operations across the world. Major operators have built up their know-how to confront them as best as possible, and at the international level, cooperative efforts have led to the formulation of elaborate guidance documents, primarily in the context of IAEA[4] and OECD/NEA.[5]

This international guidance reflects a consensus on best practices observed in the field, organized against the backdrop of the structure of safety requirements provided by IAEA nuclear safety principles and standards. It thus provides a logical and convincing doctrinal framework which appears to offer readymade practical answers to expected decommissioning problematics. However, while these documents emphasize the importance of cultural and organizational factors, they fail to provide deep insights into the managerial mechanisms that lie at the root of problems, alongside more visible technical issues. They do not warn that the effectiveness of recommended procedures depends on mostly invisible organizational limits (Farjoun and Starbuck 2007; Oliver et al. 2017) which are specific to each entity. They offer static solutions, whereas organizational dynamics (Melão and Pidd 2000) are constantly changing the background against which they need to be implemented, requiring sufficient agency to make managerial adjustments to ensure, for example, that every task always makes sense for those responsible for their execution. Therefore, managing decommissioning projects requires more than knowing and executing a pre-set doctrine, technically and economically applied to a specific project.

---

[4] IAEA safety Standards, decommissioning of facilities, General safety Requirements, Part 6.

[5] OECD NEA Report Decommissioning Funding: Ethics, Implementation, Uncertainties ISBN 92-64-02312-7.

OECD NEA Report Selecting Strategies for the Decommissioning of Nuclear Facilities ISBN 92-64-02305-4.

A similar analysis has been made in the context of the more generic problem of nuclear safety management. The international nuclear safety community has recognized that managing safety requires leadership and has developed innovative requirements in this respect.[6] The question arises however of how best to train managers to effectively exercise leadership for safety, beyond the mostly performative approach of learning about those regulatory requirements.

9.3 Management and Leadership for Safety Education

This was the goal of the European Leadership for Safety Education[7] (ELSE) Project, funded by the European Union,[8] which developed a master-level professional course and related University Diploma, based on the latest findings in management and social sciences, and on relevant industry expertise. The ELSE syllabus[9] allows trainees to explore, and then practice through a tutored personal project at their workplace, several complementary themes. Supported by a multi-national and multi-disciplinary pedagogical team, this course contributes to helping manager-trainees reflect and think critically on how best to confront ongoing challenges of effective safety management in a collective high-risk and highly regulated workplace environment characterized by a high level of uncertainty. This list of themes includes the relationships between risk and safety: going beyond technical matters and probabilities, to understand the role of risk perception, safety culture, and safety climate; the high-reliability organizations (HRO) (La Porte 1996; Roberts 1990; Weick and Sutcliffe 2001) approaches; the role of organizational limits and rule-making methods (Grote et al. 2009), and their influence on mindfulness (Kudesia 2019; Weick and Sutcliffe 2006) and managing uncertainty; organizational dynamics and leadership as a process (imbedded in complex collective relationships, rather than seen as an individual skill) (Fischer et al. 2017) and underlying mechanisms that relate to safety enhancement (sense-giving and sense-making, mindfulness, agency in a highly regulated workplace); learning and knowledge management processes (Alavi and Leidner 2001; Nesheim and Gressgård 2014; Nonaka 1994).

Based on this first successful experience and acknowledging that most of the training programmes related to nuclear decommissioning focus on technical matters, a similar approach is now under development to address the specific challenges of

---

[6] IAEA safety Standards, leadership and management for safety, General safety Requirements Part 2.
[7] https://univ-cotedazur.eu/european-leadership-for-safety-education/else-training.
[8] Grant Contract number (INSC/2019/401-273).
[9] https://univ-cotedazur.eu/medias/fichier/else-syllabus-septembre-2023-2-_1701260705940-pdf?ID_FICHE=1101530&INLINE=FALSE.

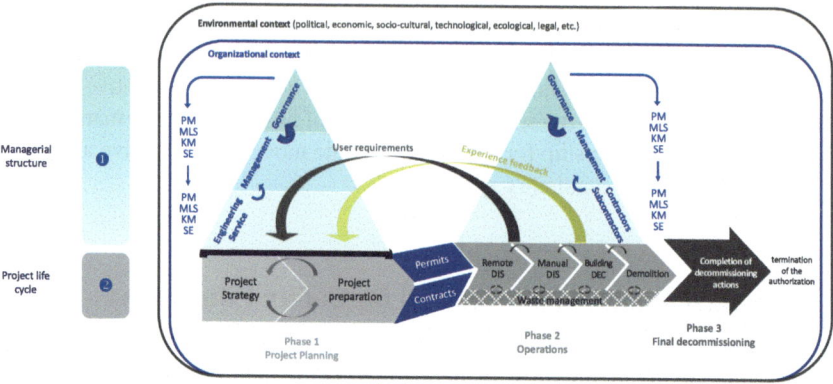

Fig. 9.1 Nuclear decommissioning project model (DIS: Dismantling, DEC: Decontamination)

managing complex nuclear decommissioning projects. Similar to ELSE, the Decommissioning Management and Leadership for Safety Education (DMaLSE) project,[10] co-funded by the EU,[11] aims to create a master-level training programme based on the state-of-the-art in management science and organizational best practices.

Through an integrative review of both academic and grey literature, along with interviews with nuclear industry experts, we have identified the key themes that should be incorporated into such a training programme to address the unique challenges associated with nuclear decommissioning projects. Besides management and leadership for safety, which includes implementing integrated management systems, a strong focus on managing complex projects (Daniel and Daniel 2018; Geraldi et al. 2011) needs to be integrated. This theme encompasses project governance, project architecture, project and programme management, human resources management, and sustainability and circularity. Related to these two main themes, particular attention needs to be paid to the areas of knowledge management and stakeholder engagement (Branko et al. 2022; Cotton 2009; Fahlquist and Roeser 2015). Furthermore, considering the current development pace of digital technologies in support of nuclear decommissioning projects (NDP), a focus needs also to be made on the role, usage and limits of such tools in these projects (Patterson et al. 2016). Finally, the interactions between the various stages of the decommissioning project and waste management issues will be explored.

To have a global representation of how these different themes are incorporated in NPD, a global model has been developed, represented by Fig. 9.1.

This model presents the three main phases of NDP for nuclear facilities (project planning, project execution and final decommissioning), the managerial structure (governance, project management and other parties involved) of the first two phases,

---

[10] https://univ-cotedazur.eu/decommissioning-management-and-leadership-for-safety-education-dmalse.

[11] Grant Contract number INSC/2022/432-533.

and the informational links between each phase. The managerial structure (1) oversees the implementation of the directives and recommendations of the IAEA and EURATOM in terms of project management (PM), management and leadership for safety (MLS), knowledge management (KM) and stakeholders engagement (SE) over the project life cycle (2). The stages of a NDP project include waste management as well as the design and use of digital technologies, which facilitate the step-by-step planning of deconstruction operations, the operational training of operators, by providing detailed three-dimensional representations of buildings and equipment. Numeric models also facilitate the design and intervention of robotics solutions for treating areas where contamination levels are not compatible with direct human interventions.

A workshop with 22 nuclear industry experts has been organized during an international conference dedicated to nuclear decommissioning to present these themes and the model. Experts have been asked to share their experiences regarding the managerial challenges related to the themes. Some of their answers are presented in Table 9.1 as examples of their contributions. These different testimonies show the diversity of the challenges related to the key themes as well as their interconnection, as numerous contributions made for one theme refer to another.

This exercise helped us refine the analytical grid that will be used to study several cases of ongoing NDPs, at different stages and in different countries with different doctrines on nuclear decommissioning organization: in some countries, the licensed nuclear operator remains responsible for the decommissioning operations until all relevant radioactive materials have been removed from the site, or treated on site so as to ensure the public's safety after the end of the licensing phase. In other countries, the facilities are transferred at the end of their operational life to a national decommissioning authority which then takes responsibility for NDP operations and radioactive waste management. In the first model, the benefit is the availability of staff who know perfectly the site and its buildings and equipment, the downside being potentially a limited experience in decommissioning activities, while the second model proposes the reverse conditions. Both models present significant challenges, and the case studies proposed by the DMaLSE Project aim to provide an in-depth understanding of the concrete organizational situations and the context in which they are embedded, allowing individuals participating later in the training to learn how to overcome the inevitable multiple challenges and uncertainties which characterize NDP activities. Various data sources will be mobilized for the elaboration of case studies, including semi-structured interviews and internal documentation. The interviews will be done with key individuals and stakeholders involved in and/or impacted by the selected NDPs (including safety authorities, TSOs, representatives of civil society and local communities, etc.). Figure 9.2 illustrates the key generic questions for each of the main themes.

Table 9.1 Examples of contributions from nuclear industry experts for each of the key topics

Key topics	Examples of contributions
Managing complex projects	(Time–cost) balance—safety first; Clear roles and responsibilities; Contracting issues, Control of assumptions (cross-projects); Coordination of all processes and projects in org: "helicopter view"; Interdependencies; Missing acceptance of Agile PM Methods; Motivation of young generation; Multidisciplinary approaches required/Sequence and time issues for activities application; Priorities/Stakeholder engagement/Enabling factors/Critical assets; Prioritizing of work/Lack of Funding; Risk of changing scope; Time organization; Unknown risk
Management & Leadership for safety	Better understanding and prioritizing of safety within leadership; Budgetary constraints; Building the proper management structure to lead the project; Open communication; Decision-making/strategic level/top-down process and bottom-top process; Good processes for addressing "toxic behaviour" in the culture; Have time/competence in feedback culture; to address wanted and unwanted behaviour; Is cooled by managerial attitude; Leading by example; Missing leadership commitment; Quickly changing environment; Safety culture
Knowledge management	Daily struggle with hierarchy and old approaches to education; Distributed systems; violation of single point of truth; Distribution of information; Diverse knowledge management systems (databases); Exaggerated access control; Generation change; Intellectual property issues; Know-how/expert judgement/data management; Knowledge transfer; Legacy data—repository Long-term storage of data—a challenge; Loss of knowledge; Missing knowledge culture; Need available tools to collect information; Shortage of experts/Building next generation of experts in a short time; Social aspects are not taken into account; Tacit knowledge
Stakeholder engagement	Communication—language—level of detail; Decision approach (lack of consensus); Dialogue with stakeholders; Disbelief + mistrust; How to identify the main stakeholders? Low level of knowledge of public to make decisions; Multi-stakeholder/multi-actors/public expectations; Over-selling/under-estimation/over-confident; Planification; Public perception; Regular meetings; Stakeholders not presenting their immediate and long-term issues

(continued)

Table 9.1 (continued)

Key topics	Examples of contributions
Waste management	Circular economy and cost from waste owners' perspective; Contradictory optimization aims; Interdependencies problematic waste; Is reprocessing of spent fuel better than Deep Geological Repository (DGR) ?; Language barrier Packaging constraints; Recycle of waste; Stakeholder opposition to the release of cleared waste; Too strict clearance/release limits; Tracking and bookkeeping is a challenge; Waste characterization; Waste management: harmonization
Design and usage of digital technologies	Acceptance inside organization; Acceptance by regulator; Access to relevant data; Administration slows things down and limits possibilities; AI—Support decision/databases management/Inter-operability for multi-stakeholders; Closed technological platforms; Cost–benefit analyse for robotics applications; Cyber security; Diverse terminology; Documenting dismantling work taking > 100 photos daily; Identification of limits for use of robots, AI; Interconnection of data; Labour unions; Licenses, access to software; Missing interoperability; Paper-centric documentation approach; Practical effective use to be proven; Proprietary formats; Standardization is missing; Validation of models without "real-life" version; Wifi issues

Fig. 9.2 DMaLSE main topics of interest and their key generic questions

9.4 Conclusion

The challenges associated with nuclear decommissioning extend beyond technical aspects, involving intricate managerial complexities and organizational dynamics. NDP present unique challenges, such as a paradoxical relationship with time, sociocultural burdens related to radiation exposure risk, and the inherent uncertainty and complexity of the D&D projects. While international guidance documents provide a doctrinal framework based on best practices, they often lack insights into the organizational roots of challenges specific to each entity.

The Decommissioning Management and Leadership for Safety Education (DMaLSE) project, funded by the European Union, addresses this critical gap by developing a master-level training programme tailored for professionals engaged in nuclear decommissioning. Inspired by the successful European Leadership for Safety Education (ELSE) Project, the DMaLSE initiative adopts a holistic approach, integrating the latest findings in management science and organizational best practices. While most existing courses focus either on technical problematics, on safety rules and practices, or on regulatory and legal aspects of nuclear decommissioning, the DMaLSE approach puts training participants in the centre of managerial problematics where all these issues are intricated, preparing them to future real-life situations. Themes such as management and leadership for safety, complex project management, knowledge management, stakeholder engagement, and the role of digital technologies are incorporated to equip professionals with the skills needed for the complexities of nuclear decommissioning.

Effective management and leadership are indispensable for the success of nuclear decommissioning projects. The DMaLSE project emerges as a significant initiative, addressing the managerial complexities inherent in these projects and paving the way for a new generation of professionals equipped to tackle the multifaceted challenges of nuclear decommissioning. Continuous refinement, research, and collaboration will create a safer and more efficient future for the decommissioning of nuclear facilities worldwide. Through these efforts, we strive to ensure the success and sustainability of nuclear decommissioning projects, fostering a culture of safety and excellence in the industry.

References

M. Alavi, D.E. Leidner, Review: knowledge management and knowledge management systems: conceptual foundations and research issues. MIS q. **25**(1), 107–136 (2001). https://doi.org/10.2307/3250961

K. Branko, B. Paul, F. Simon, P. Alan, Z. Ming, Y. Tamara, W. Michael, P. Stephane, B. Tine, B. Marko, Demonstrating the use of a framework for risk-informed decisions with stakeholder engagement through case studies for NORM and nuclear legacy sites. J. Radiol. Prot. **42**(2), 020504 (2022). https://doi.org/10.1088/1361-6498/ac5816

M. Cotton, Ethical assessment in radioactive waste management: a proposed reflective equilibrium-based deliberative approach. J. Risk Res. **12**(5), 603–618 (2009). https://doi.org/10.1080/13669870802519455

P.A. Daniel, C. Daniel, Complexity, uncertainty and mental models: from a paradigm of regulation to a paradigm of emergence in project management. Int. J. Project Manage. **36**(1), 184–197 (2018)

J.N. Fahlquist, S. Roeser, Nuclear energy, responsible risk communication and moral emotions: a three level framework. J. Risk Res. **18**(3), 333–346 (2015). https://doi.org/10.1080/13669877.2014.940594

M. Farjoun, W.H. Starbuck, Organizing at and beyond the limits. Organ. Stud. **28**(4), 541–566 (2007). https://doi.org/10.1177/0170840607076584

T. Fischer, J. Dietz, J. Antonakis, Leadership process models: a review and synthesis. J. Manag. **43**(6), 1726–1753 (2017). https://doi.org/10.1177/0149206316682830

J. Geraldi, H. Maylor, T. Williams, Now, let's make it really complex (complicated). Int. J. Oper. Prod. Manag. **31**(9), 966–990 (2011). https://doi.org/10.1108/01443571111165848

G. Grote, J.C. Weichbrodt, H. Günter, E. Zala-Mezö, B. Künzle, Coordination in high-risk organizations: the need for flexible routines. Cogn. Technol. Work **11**(1), 17–27 (2009). https://doi.org/10.1007/s10111-008-0119-y

R.S. Kudesia, Mindfulness as metacognitive practice. Acad. Manag. Rev. **44**(2), 405–423 (2019). https://doi.org/10.5465/amr.2015.0333

T.R. La Porte, High reliability organizations: unlikely, demanding and at risk. J. Contingencies Cris. Manag. **4**(2), 60–71 (1996)

N. Melão, M. Pidd, A conceptual framework for understanding business processes and business process modelling. Inf. Syst. J. **10**(2), 105–129 (2000)

T. Nesheim, L.J. Gressgård, Knowledge sharing in a complex organization: antecedents and safety effects. Safety Sci. **62**, 28–36 (2014). https://doi.org/10.1016/j.ssci.2013.07.018

I. Nonaka, A dynamic theory of organizational knowledge creation. Organization Sci. **5**(1), 14–37 (1994). http://www.jstor.org/stable/2635068

N. Oliver, T. Calvard, K. Potočnik, Cognition, technology, and organizational limits: lessons from the Air France 447 disaster. Organ. Sci. **28**(4), 729–743 (2017). https://doi.org/10.1287/orsc.2017.1138

E.A. Patterson, R.J. Taylor, M. Bankhead, A framework for an integrated nuclear digital environment. Prog. Nucl. Energy **87**, 97–103 (2016)

K.H. Roberts, Managing high reliability organizations. Calif. Manage. Rev. **32**(4), 101–113 (1990). https://doi.org/10.2307/41166631

K.E. Weick, K.M. Sutcliffe, *Managing the unexpected*, vol. 9 (Jossey-Bass, San Francisco, 2001)

K.E. Weick, K.M. Sutcliffe, Mindfulness and the quality of organizational attention. Organ. Sci. **17**(4), 514–524 (2006). https://doi.org/10.1287/orsc.1060.0196

Open Access This chapter is licensed under the terms of the Creative Commons Attribution 4.0 International License (http://creativecommons.org/licenses/by/4.0/), which permits use, sharing, adaptation, distribution and reproduction in any medium or format, as long as you give appropriate credit to the original author(s) and the source, provide a link to the Creative Commons license and indicate if changes were made.

The images or other third party material in this chapter are included in the chapter's Creative Commons license, unless indicated otherwise in a credit line to the material. If material is not included in the chapter's Creative Commons license and your intended use is not permitted by statutory regulation or exceeds the permitted use, you will need to obtain permission directly from the copyright holder.

Chapter 10
The Language of Transitions: Navigating Innovation, Decline and Renewal

Eric Marsden and Mathilde Bourrier

Abstract This concluding chapter reflects on the connotations of the vocabulary and linguistic patterns used by scholars, policymakers, and practitioners when discussing transitions and discontinuations. Through the lens of history of Royaumont Abbey, where the workshop that gave rise to the current volume was held, a site that has witnessed cycles of innovation, decline, and renewal, we examine the political and symbolic implications of terms commonly used to discuss phase-out trajectories. The language often carries negative undertones but also includes references to renewal and rebirth, highlighting the complex and multifaceted nature of transitions.

Keywords Language · Vocabulary · Transitions · History · Lifecycle phases

The workshop which gave rise to this book was organized at Royaumont, a former Cistercian abbey located slightly to the north of Paris, which provided a delightful contemplative backdrop to the discussions between the contributing authors and the members of NeTWork. Royaumont was built in a seven-year period around 1230 with the support of French king Louis IX (known as « Saint Louis » or the "monk king", who also commissioned famous buildings such as the Sainte Chapelle in Paris). Its rapid construction was an impressive feat, which is reported to have consumed two thirds of the budget of the monarchy over the period. The abbey grew to be very powerful, housing almost 150 monks who contributed to the dissemination of knowledge and to the production of the *Speculum Majus* encyclopaedia. The abbey's influence waned during the 100-year war, as it suffered from pillaging and from a loss of legitimacy due to decadent practices adopted by the monks in the fifteenth century. After the French revolution, which dismantled religious orders, the buildings at Royaumont were sold to an entrepreneur who established an industrial bleachery and a cotton mill on the grounds. Stones from the abbey church were repurposed to build housing for the 300 workers on site. Two water wheels were installed to power the machines thanks to the diversion of a nearby river; they were replaced thirty years

---

E. Marsden (✉) · M. Bourrier
Foundation for an Industrial Safety Culture, Toulouse, France
e-mail: eric.marsden@foncsi.org

© The Author(s) 2025
M. Bourrier (ed.), *Decommissioning Aging Installations and Declining Technologies*,
SpringerBriefs in Safety Management, https://doi.org/10.1007/978-3-031-88369-9_10

later by cutting-edge steam engines. This industrial activity later entered a phase of decline due to competition from other regions of France, and the activity shut down in 1863. The abbey was then purchased by a rich family and became their summer residence. During the First World War, the abbey hosted a hospice for injured soldiers run by Scottish suffragettes, where innovative medical practices for the treatment of gas gangrene were developed. Royaumont then became the home to a foundation for the performing arts, with a secondary function as a conference centre.

Royaumont's rich history, encompassing phases of life dedicated successively to religious, industrial, cultural, and commercial activities, is a perfect illustration of the cycles of innovation, decline, renewal, and reuse that were the subject of the workshop. The industrial and social transitions that the abbey lived through—whether triggered by abrupt, and even revolutionary, changes in power, or slow processes that played out over long periods of decline and neglect, mirror the different transitions discussed by the participants. The historical role of monasteries and abbeys in archiving knowledge and copying manuscripts to allow the dissemination of new ideas resonates with the emphasis given in several contributions in this volume to the importance of archival and museographic work in maintaining a record of past activities. The religious dimension of the abbey's history was appropriate, given that both transition theories and the religious rituals associated with end-of-life and renewal, deal with the end of one state and the careful preparation for a new beginning, and build on shared concepts of purification and cleansing, transformation and rebirth, ritual and ceremony, respect for both the past and the future.

In this concluding chapter, we reflect on the connotations implied by the vocabulary and patterns of language used by scholars, policymakers and practitioners to refer to the different forms of discontinuation or phase-out trajectories discussed in this volume. The terminology used to describe technologies is rarely politically neutral. Consider for example the phase-out of pesticides discussed by Turnheim et al. in this volume, where the agri-food lobby has attempted to replace the words "pesticide" and "insecticide", whose Latin roots refer to the death of living creatures, by the term "plant protection" or "phytosanitary" products.

The language of discontinuation and transition often carries negative or ominous undertones, frequently using the Latin prefix de- ("down from") or dis- ("opposite of, apart"). Consider for example the terminology used by regulators: decommissioning (the process that leads to the irreversible closure of a facility and the termination of the operating license), deconstruction of facilities and equipment, their depollution, the decontamination, dismantling, disposal of discarded waste. Researchers in the science and technology studies field write of destabilization, decline, divestment (a term associated with social campaigns that attempt to delegitimize a type of activity), discontinuation, dissolution, disinvestment, and disnovation. Some researchers highlight innovation through withdrawal (Goulet and Vinck 2023), referring for example to "OGM-free" and "preservative-free" foods. Ecologists debate the importance of "degrowth" as a response to the depletion of natural resources and destabilization of the climate.

However, other terminology includes more positive references to eternal life and salvation, emphasizing the renewal aspects of transitions and the "salvage" value of assets, in particular through the Latin prefix "re-" for again, once more. We talk of the environmental "remediation" and the "redevelopment" of brownfield sites, which are increasingly seen as opportunities to benefit from and "repurpose" land which is well-connected to infrastructures, rather than as burdensome legacies from a failed industrial past. The operators of a mine must in principle "rehabilitate" it once extraction has ended, for example using topsoil replenishment and reforestation, and the area is "reclaimed", "restored", "renatured", and "revitalized". Industrial facilities undergo "revamping" to replace declining technologies with newer ones and refurbish or repair equipment. A technology may be "rebranded" to attempt to project a more positive image (for example, natural gas presented as a green transition fuel rather than as a fossil fuel). The Schumpeterian link between destruction and creation is emphasized by terminology such as transition, transformation, substitution, and more recently "out-innovation" or "exnovation", terms proposed to refer to the flip side of innovation (David and Gross 2019). Politicians refer to "sunset industries" such as coal mining and textile manufacturing in Europe, with a symbolism that implicitly includes the coming sunrise. The corporate world also favours positive language that emphasizes continuation of activity and the associated benefits. Firms' websites emphasize the use of sophisticated technologies for "reverse installation" and "end-of-life management", which is becoming a recognized industry sector which puts forward its environmental credentials, recovering and "recycling" waste products in the "circular economy". A company that provides recycling and waste management services for the nuclear industry is called "Cyclife". Creative individuals "upcycle" waste materials to give them a new life.

Other terminology used by practitioners is more pragmatic. In the nuclear energy sector, "longer-term operation" is an important area of focus in allowing certain power plants to run for longer than their original design age, with significant economic benefits. In the offshore oil and gas sector, consulting firms emphasize the economic opportunities associated with efficient end-of-life management and "decommissioning excellence", referring to the similarities to the "pioneering spirit" present during the initial expansion of the drilling industry. Some firms specialize in "late-life asset management", optimizing the "late-life operations" of a "mature" or "declining" asset before "cessation of production" (COP). The extractive industries use quite brutal terminology, such as the "plugging and abandonment" (P&A) of oil wells; "orphaned" mines constitute a significant safety and environmental hazard in several countries.

Some new terms are used to emphasize specific activities or important actors in a discontinuation transition: Stegmaier (this volume) refers to "discontinuation entrepreneurs", people who use the legal framework in new ways to political ends, and to "aftercare" to underscore the need for work looking after a system or technology once it is on the decline.

The transitions and discontinuation activities discussed during the workshop are invariably significant, complex and problematic. As mentioned in the introductory chapter, they are sometimes treated as taboo topics by practitioners and decision-makers, as sad stories of suffering. However, both the rich multi-century history of Royaumont abbey with its multiple cycles of activity, and the fruitful workshop discussions that have informed the contributions in the present volume, bear witness to the relevance of making the effort to overcome the taboo and use past experience better to navigate future transitions, making them safer and less problematic.

References

M. David, M. Gross, Futurizing politics and the sustainability of real-world experiments: what role for innovation and exnovation in the German energy transition? Sustain. Sci. **14** (2019). https://doi.org/10.1007/s11625-019-00681-0

F. Goulet, D. Vinck, *New Horizons for Innovation Studies—Doing Without, Doing With Less* (Edward Elgar Publishing, 2023) ISBN 9781803925547

Open Access This chapter is licensed under the terms of the Creative Commons Attribution 4.0 International License (http://creativecommons.org/licenses/by/4.0/), which permits use, sharing, adaptation, distribution and reproduction in any medium or format, as long as you give appropriate credit to the original author(s) and the source, provide a link to the Creative Commons license and indicate if changes were made.

The images or other third party material in this chapter are included in the chapter's Creative Commons license, unless indicated otherwise in a credit line to the material. If material is not included in the chapter's Creative Commons license and your intended use is not permitted by statutory regulation or exceeds the permitted use, you will need to obtain permission directly from the copyright holder.

The manufacturer's authorised representative in the EU is Springer Nature Customer Service Centre GmbH, Europaplatz 3, 69115 Heidelberg, Germany. If you have any concerns regarding our products, please contact ProductSafety@springernature.com

Printed and bound by CPI Group (UK) Ltd, Croydon, CR0 4YY

23/03/2026

02076360-0010